Optical Mineralogy

Pramod K. Verma

Optical Mineralogy

Pramod K. Verma
Department of Geology
University of Delhi
Delhi, India

ISBN 978-3-031-40764-2   ISBN 978-3-031-40765-9 (eBook)
https://doi.org/10.1007/978-3-031-40765-9

Jointly published with Ane Books Pvt. Ltd.
In addition to this printed edition, there is a local printed edition of this work available via Ane Books in South Asia
(India, Pakistan, Sri Lanka, Bangladesh, Nepal and Bhutan) and Africa (all countries in the African subcontinent).
ISBN of the Co-Publisher's edition: 978-93-80156-08-8

1st edition: © Author 2014
© The Author(s) 2023

This work is subject to copyright. All rights are solely and exclusively licensed by the Publisher, whether the
whole or part of the material is concerned, specifically the rights of reprinting, reuse of illustrations, recitation,
broadcasting, reproduction on microfilms or in any other physical way, and transmission or information storage
and retrieval, electronic adaptation, computer software, or by similar or dissimilar methodology now known or
hereafter developed.
The use of general descriptive names, registered names, trademarks, service marks, etc. in this publication does
not imply, even in the absence of a specific statement, that such names are exempt from the relevant protective
laws and regulations and therefore free for general use.
The publishers, the authors, and the editors are safe to assume that the advice and information in this book are
believed to be true and accurate at the date of publication. Neither the publishers nor the authors or the editors
give a warranty, expressed or implied, with respect to the material contained herein or for any errors or omissions
that may have been made. The publishers remain neutral with regard to jurisdictional claims in published maps
and institutional affiliations.

This Springer imprint is published by the registered company Springer Nature Switzerland AG
The registered company address is: Gewerbestrasse 11, 6330 Cham, Switzerland

Preface

Over the last several decades, the increasing use of microprobe techniques, such as electron probe, ion probe, raman laser probe, etc., has been replacing optical mineralogy techniques as the method of choice for investigation. However, the need to understand and resolve rock textures under microscopes has resulted in a renewed interest in optical mineralogy. In particular, the thermobarometry of inclusions in phenocrysts and porphyroblasts requires a thorough microscopic examination of rocks. Consequently, after a gap of nearly 30 years or so, when such classics as Roger and Kerr (later by Kerr), Winchell, and Whalstorm were the common textbooks, new textbooks in this field have come out. Important among them are those written by Phillips and Griffen, Gribble and Hall, Deer, Howie & Zussmann (student addition), and by Nesse.

These books cover the subject matter in an exceedingly comprehensive manner for selected topics. Students of geology face a unique problem as often pointed out. They prefer a book of the right size, with the right details, that can be brought to a classroom, a laboratory, as well as on field trips across difficult terrains. They would like the book to reflect the new applications, but without losing on the details of the fundamental principles. To meet this varied and diverse requirement is not an easy task though I have, through this book, attempted to fill this important void. It is the culmination of nearly four decades of my teaching career. The book reflects my approach to teaching Optical Mineralogy, an approach that has involved continuous experimentation on the teaching methodology, and feedback from many brilliant students. Following several authors, like Kerr and Nesse, this book too is divided into two parts. Part I deals with the theory and techniques, and, Part II provides a description of optical properties of common minerals. Determinative tables assembled as Appendices should help students in identification when tutors are not around. While it is assumed that students have some basic knowledge of optical mineralogy, as provided in standard mineralogy textbooks such as Dana's Manual of Mineral Sciences, 23rd edition (Cornelius Klein and Elizabeth Dutrow: John Wiley), some important requisite back material on Optics has also been included as Technical Text Boxes. At the end of Part I, a chapter has been added to acquaint freshman students about the applications of optical mineralogy which they might employ in their career.

A number of my former students have earned reputation country-wide for their skill in handling rocks and minerals. Every one of them has been of great help to me in writing this book. In particular, I would like to mention Naresh Pant, Amitav Kundu,

Sarbari Nag, Preeti Singh, and Balakrishnan for their constant cooperation and words of encouragement. Kulanand Kandwal, the Museum Curator of Geology Department at Delhi University, has been assisting me in arranging laboratory materials for my undergraduate and graduate courses for the past 25 years. He along with K.K. Singh was always eager to help, and I am indebted to them for their co-operation. Thanks are also due to Prof. A.M. Bhola, Head of the Geology Department for allowing me to use the facilities. To those who provided material and accorded permission to their use in the book, I shall remain obliged. These friends and agencies are acknowledged in the body of the text. I have often received instant guidance from the notes given in Wikepedia on the internet; I am sure most students also fall back on this and other sources from internet.

I am indebted to the undergraduate students of Geology at Delhi University, Class of 2009, particularly the group at Ram Lal Anand College, for their strong insistence that I bring out this book.

I shall be failing in my duties if I don't express sense of gratitude towards Sh. Sunil Saxena and Sh. Jai Raj Kapoor of Ane Books Pvt. Ltd. who pulled me out of their pool of acquaintance, and put me on the path that led to the completion of the book. Their editor Nishant Saini, and their master graphic expert R.K. Majumdar have done a wonderful job. I must express my thanks to Cathy Giacari of Taylor and Francis for devoting her expertise in bringing this book to its final shape. In particular, I would like to thank the future readers of this book in advance for any feedback on the book.

My family members pitched in for me during the two years of its preparation. Son Amit was always helpful with his comments, daughter Abhilasha and daughter-in-law Tejal cheered me up and provided frequent internet accesses, as well as help with their knowledge of MS Office, wife Surinder's often repeated challenge that I would never finish this work, did the trick!

Author

Abbreviations and Symbols

Å	Ångstrom units (10^{-10}m)
a	cell edge in the x-direction
at. per cent	atoms per cent, percentage on atomic basis
apfu	atoms per formula unit
b	cell edge in the y-direction
Bx_a	acute bisectrix
c	cell edge in the z-direction
d	interplanar spacing
Fe*	total $Fe^{2+} + Fe^{3+}$
M1, M2 etc.	Metal sites in a crystal structure
n	refractive index of glass and isometric minerals
OAP	optic axial plane
pfu	per formula unit
ppm	parts per million
ppb	parts per billion
R	metal ions
RE, REE	rare earth, rare earth elements
r < v	dispersion; indicating
wt per cent	weight per cent, percentage on a weight basis
x, y, z	crystallographic directions
$X_{Fetotal}$	Total iron as mole fraction
Z	number of formula units per unit cell

2V	optic axial angle
2Vá	optic axial angle with á optic direction as bisectrix
2Vã	optic axial angle with ã optic direction as bisectrix
á, â, ã	optic directions, refractive indices along these directions; á < â < ã by convention
ä	birefringence
å	extraordinary ray, refractive index of extraordinary ray
ù	ordinary ray, refractive index of ordinary ray

Contents

Preface   v

Abbreviations and Symbols   vii

> **Principles and Techniques of Optical Mineralogy**   **Part I**

1. Properties of Light   3

- ☐ Introduction   4
- ☐ Dual Nature of Light   4
- ☐ Light as Transverse Waves   7
- ☐ Wave Surfaces   8
- ☐ Optical Phenomena   10
- ☐ Polarization of Light   12
- ☐ Double Refraction   13
- ☐ Interference of Light   14
  - *Summary*   17
  - *Important Terms*   18
  - *Questions*   18

2. Sample Preparation for Transmitted Microscopy   19

- ☐ Introduction   20
- ☐ The Polarizing Microscope – Primary Equipment for Optical Mineralogy   20
- ☐ The Purpose of Sample Preparations   21
- ☐ Nature of Samples   21
- ☐ Size of Samples   22
- ☐ Thickness of a Sample and the Concept of Transparency   24
- ☐ Polishing and Mounting Rock Chips   25
  - *Summary*   29
  - *Important Terms*   30
  - *Questions*   30

ix

3. Refractometry 31

- Introduction 32
- Relief 32
- Becke Line 33
- Oblique Illumination Method 40
- Refractometers 41
  - *Summary* 45
  - *Important Terms* 45
  - *Questions* 46

4. Optical Crystallography 47

- Introduction 48
- Isotropic Optics 48
- Uniaxial Optics 50
- Biaxial Optics 56
  - *Summary* 60
  - *Important Terms* 61
  - *Questions* 61

5. The Polarizing Microscope 63

- Introduction 64
- The Compound Microscope 64
- The Polarizing Microscope 70
- Use and Care of Polarizing Microscopes 76
  - *Summary* 81
  - *Important Terms* 81
  - *Questions* 82

6. Microscopic Examination of Minerals I: Orthoscopic Condition 83

- Introduction 84
- Examination in Ordinary Light 84
- Under Plane Polarized Light 92
- Crossed Polars 93
  - *Summary* 102
  - *Important Terms* 103
  - *Questions* 103

7. Microscopic Examination of Minerals II — **105**

- Examination in Crossed Polars (Contd.) — 106
- Measurement of Extinction Angle — 112
- Twinning — 112
- Zoning and Overgrowth — 116
- Optical Anomalies — 118
- Properties Requiring the Use of Compensators — 119
  - *Summary* — 126
  - *Important Terms* — 127
  - *Questions* — 127

8. Microscopic Examination of Minerals III: Conoscopic Condition — **129**

- Introduction — 130
- The Microscope as a Conoscope — 130
- Uniaxial Crystals — 132
- Biaxial Crystals — 134
- Dispersion — 140
- Optic Sign Determination — 144
  - *Summary* — 146
  - *Important Terms* — 147
  - *Questions* — 148

9. Reorienting Techniques — **149**

- Introduction — 150
- Spindle Stage — 151
- Universal Stage — 158
- Fundamental Principle — 163
  - *Summary* — 173
  - *Important Terms* — 173
  - *Questions* — 174

10. New Frontiers in Microscopy — **175**

- Introduction — 176
- Fluid Inclusion Studies — 176

☐ Image Analysis System	177
☐ Transmission Electron Microscope (TEM)	182
☐ Scanning Electron Microscope (SEM)	184
☐ Atomic Force Microscopy (AFM)	188
• *Summary*	190
• *Important Terms*	191
• *Questions*	191

Systematic Description of Common Rock Forming Minerals **Part II**

11. Nesosilicates — 195

☐ Olivine	196
☐ Monticellite	199
☐ Garnet	200
☐ Vesuvianite (Idocrase)	202
☐ Zircon	204
☐ Andalusite	206
☐ Kyanite	208
☐ Sillimanite	210
☐ Mullite	212
☐ Staurolite	213
☐ Sphene	215
☐ Topaz	217
☐ Chloritoid	219

12. Sorrosilicates and Cyclosilicates — 221

☐ Zoisite	222
☐ Clinozoisite - Epidote Series	224
☐ Allanite (Orthite)	227
☐ Piemontite	229
☐ Pumpellyite	230
☐ Lawsonite	232
☐ Beryl	233
☐ Tourmaline	235
☐ Cordierite	237

xiii

13. Inosilicates — 241

- Orthopyroxenes Series — 242
- Pigeonite — 245
- Diopside - Hedenbergite Series — 246
- Augite — 249
- Omphacite — 252
- Jadeite — 254
- Aegirine to Aegirine - Augite — 255
- Wollastonite — 257
- Sapphirine — 259
- Anthophyllite — 261
- Cummingtonite - Grunerite — 263
- Tremolite - Actinolite Series — 264
- Calcic Amphibole ("Common" - Hornblende) — 267
- Glaucophane — 271
- Riebeckite — 274

14. Phyllosilicates — 277

- Serpentine — 278
- Talc — 280
- Muscovite — 282
- Paragonite — 284
- Biotite — 286
- Chlorite — 289
- Gluconite — 292
- Apophyllite — 294
- Stilpnomelane — 295
- Prehnite — 297

15. Tektosilicates — 299

SILICA GROUP

- α-Quartz — 300
- α-Tridymite — 303
- α-Cristobalite — 305
- Coesite — 306

ALKALI FELDSPAR GROUP

- The Sanidine - High Albite Series — 308
- The Orthoclase - Low Albite Series — 311
- The Microcline - Low Albite Series — 314
- Plagioclase Feldspar — 317

NEPHELINE GROUP

- Leucite — 326
- Sodalite — 327
- Scapolite — 328

ZEOLITE GROUP

- Natrolite — 330
- Chabazite — 331
- Heulandite — 332
- Stilbite — 334
- Laumontite — 335

16. Non-silicates — 337

- Corundum — 338
- Rutile — 339
- Spinel — 341
- Perovskite — 343
- Barite — 344
- Gypsum — 345
- Anhydrite — 347
- Calcite — 348
- Dolomite — 351
- Aragonite — 352
- Apatite — 354
- Monazite — 355
- Xenotime — 357
- Fluorite — 358
- Halite — 359

Appendices — 361

About the Author

Pramod Kumar Verma a professor (Retd.) of Geology at the University of Delhi, received his Ph.D. in Geology from Harvard University, USA, in 1973. Dr. Verma has taught undergraduate and graduate-level courses in mineralogy and physical geochemistry during the last four decades. He was primarily a metamorphic petrologist but began his research in the regional geology of metamorphic rocks in the Eastern Himalayas. In addition to the Himalayas, his area of interest was in the petrology of the planetary interior. He published several research articles in leading international journals and edited two books on Eastern Himalayan Geology. Professor Verma had held visiting assignments in USA, Germany, and Zambia. He was a fellow of the Mineralogical Society London. He also served as the president of the Earth System Sciences Section of the Indian Science Congress and the vice president of the Indian Geological Congress.

PART I

Principles and Techniques of Optical Mineralogy

1
Properties of Light

Learning Objectives

- Simple concepts of optics – A review
- Polarization of light
- Double refraction
- Interference of polarized rays

Prologue

It is a matter of common knowledge that we cannot see objects around us in darkness. Hence, light enables us to see objects. Images of objects are formed in the retina of our eyes that are finally read by our brain through a highly complicated process. Some amongst us, who are unfortunate, do not have eyes; they cannot see through the eyes, yet, can get around places through physical touch, sound etc. However, many of us require spectacles to see clearly; these consist of glass material similar to those used in your binoculars or in telescopes for grazing stars. In addition, light is utilized in pursuing scientific and technical investigations about nature of materials including minerals. Optical mineralogy combines principles of Optics and description of properties of minerals based on the application of these principles.

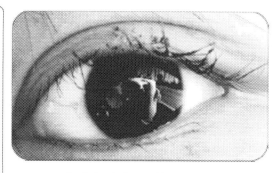

A window to the world – our eyes.

Chapter's Outline

- Dual nature of light
- Light as transverse waves
- Wave surfaces
- Optical phenomena
- Double refraction
- Interference of light

© The Author(s) 2023
P. K. Verma, *Optical Mineralogy*,
https://doi.org/10.1007/978-3-031-40765-9_1

Introduction

The optical properties of minerals are studied with the help of a polarizing microscope. In order to appreciate fully the design of this microscope as well as its limitations, and also the proper observance of optical phenomena of minerals, a background of high school optics is necessary. Most students are aware of the concepts and practices at that level, yet, it would be useful if we provide a succinct description of a few relevant parts of the optics that would be periodically used and applied throughout this book. For the confident reader our suggestion is to give this chapter a quick glance, but a few of you may require a thorough reading of this chapter or even pertinent parts of a good high school textbook. We begin with a few words about the nature of light, and then through the concept of polarization take you upto the concept of interference of light.

Dual Nature of Light

Undoubtedly, the first question is what is the nature of light? Modern science has been looking into this question since sixteenth century with many divergent views emerging from time to time. Technical Text Box 1.1 very briefly summarizes the growth of ideas about the nature of light. During the second half of the nineteenth century it became

---

Technical Text Box 1.1

UNFOLDING AND RIPENING OF THE NATURE OF LIGHT

Huygens' wave theory not only helped him in deriving laws of refraction and reflection but also explained the double refraction of light through calcite. Young's double slit camera experiment helped to consolidate the concept of wave nature of light. The middle of nineteenth century was witness to tremendous progress in this regard as Fresnel formulated equations to explain transmission and reflectance at a phase boundary of two optical media. He also showed that calcite polarized the light; Fresnel was followed by Maxwell whose set of four equations formed the basis for the electromagnetic nature of light waves.

First quarter of the twentieth century produced a series of breakthrough in our understanding of interaction between light and matter. Planck's famous equation $E = h\nu$, Einstein's explanation for photoelectric effect, Bohr's model of hydrogen atom, Compton effect, etc proved that light possesses energy and momentum. De Broglie showed in 1924 that subatomic particles are endowed with wave properties: a particle with a momentum p has an associated wavelength λ given by h/p. Experiments by Davisson, Germer, and Thomson consolidated the concept of dual nature of matter and energy. For our purpose, we may state that propagation of light, interference and diffraction are best understood in terms of waves, but photoelectric effect proves its particle nature. From another point of view, an electron is a particle for many phenomena, but as with the electron microscope it behaves like waves.

General Text Box 1.1

OPTICS OF OUR EYE

There is no relevance of optical mineralogy if one does not have eyes or healthy eyes. As shown in the diagram, the outermost part of the eye is the pupil which allows light to enter. It is situated in the centre of the eye when one is looking straight. This opening can be enlarged or shortened by means of a set of muscles called iris. Iris prevents excess light to get into the interior of the eye. The lens, situated at the back of the iris, allows the image to form on the retina. The eyes are well protected from dust etc. by eyelids and eyelashes. Blinking helps to keep cornea to receive moisture that is vitally need for its transparency. Retina is made up of rod and cone like cells. Rod-shaped cells are responsible for black and white vision whereas cone-shaped cells control color vision.

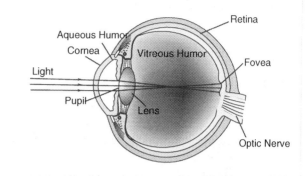

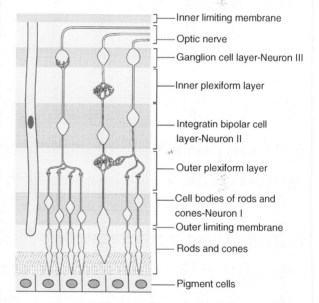

The eye lens is case 2 lens system in which the image is formed on the retina, *i.e.*, on a side opposite to that of the object. The image is smaller than the object and is inverted. In general, the object is situated at a point that is more than twice the focal length of the system. The lens system consists of three lenses that allow us to focus the image of an object on the retina. Outermost is the aqueous lens with a strong refracting power, then the eye lens, and finally, the vitreous lens, which is supposed to be main instrument for magnification. The relevant information about the object is transmitted from the retina to the human brain by means of optic nerve. The brain prepares the image as right side up and estimates about the color and the size of the object.

obvious that light is some form of radiation; implicit in this is that there are other forms of radiations. Let us now proceed to examine the place of light as a part of the known radiation spectrum which is given in Fig. 1.1. Top box of this figure represents cosmic

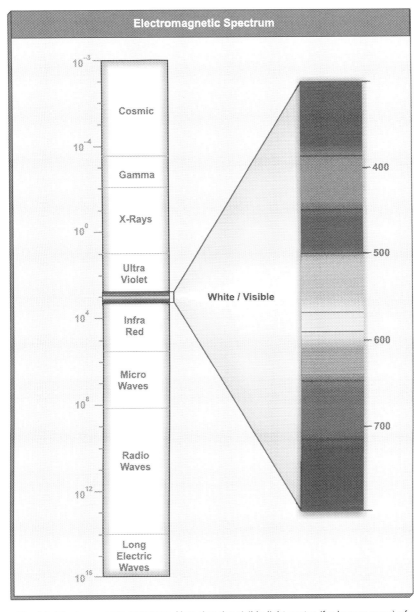

Fig. 1.1 Electromagnetic spectrum. Note that the visible light sector (for human eyes) of this spectrum occupies a very small part.

Properties of Light 7

rays that have a wavelength of the order of 10^{-8} nm. In order of increasing wavelength and downwards in the figure are gamma rays finally ending with long electrical waves. Our eyes are capable of detecting only a very small part of this spectrum, *i.e.*, approximately between 400 to 700 nm. We all know it by the name of visible spectrum. Interaction of this small part with the structure and elements of minerals is all that forms the Optical Mineralogy. Interaction of rest of the radiations with minerals is commonly covered by the Spectroscopic Studies of minerals, – topics that are beyond the scope of a beginner in mineralogy.

The light is particle and wave at the same time. We can explain some phenomena by the wave nature of light rather easily. According to Maxwell, waves propagate in 3-D space as an interaction of two vectors: (*a*) electrical, and (*b*) magnetic. Out of these two vectors, the electric vector is much stronger; hence, it is a common practice amongst physicists to carry out calculations with reference to electrical vector while dealing with the propagation of light through material. Following this practice let us imagine that the electric vector of light interacts with an atom of a given substance. Electrons in the outermost orbit of the atom receive certain acceleration because the energy of the electrical vector is now partly transferred to them. Increased momentum of the electrons causes them to become a source of radiation that is identical to the incident radiation. This is *coherent scattering of the radiation* by the electron, and results in transmission of the radiation (light). It is possible to look at the radiation-electron interaction from another angle where we may presume that a photon of the radiation collides with the electron of the atom. This results in the modification of the velocity and direction of the photon; since the energy of the photon is modified by the collision, an *incoherent radiation* is generated. A *fluorescent radiation* is also produced by the radiation-electron interaction when the incident photon is sufficiently energized to eject an electron from the inner parts of the atom. This radiation is characterized by the properties of the atom and not by the properties of the incident radiation. For the scope of the present book only the coherent scattering is relevant as it is this that causes phenomena such as reflection, refraction, diffraction etc. Other types of scattering are utilized in spectroscopic studies of minerals that you will also learn in due course.

Light as Transverse Waves

You know that there are two kinds of waves that can be generated in a given medium: (*a*) Longitudinal waves, and (*b*) Transverse waves. The longitudinal waves have vibrations along or parallel to their direction of travel whereas transverse waves are those that oscillates perpendicular to the direction it advances. Major components of a transverse wave are shown in Fig. 1.2.

The red color straight line A-B-C is a part of the light ray that is propagating from left to right. It is a result of a transverse motion of the medium; the wave motion is shown as blue line A-E-B-C. The particles of the medium at A move up, but that at B move down during the passage of successive waves. As the wave motion reaches C particles again start moving up. The points A and C are thus called points in the same phase but B is in the opposite phase. The

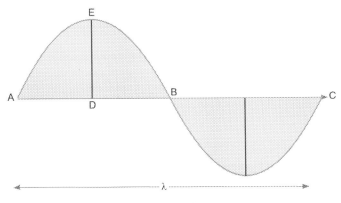

Fig. 1.2 Components of a wave.

distance D-E is called the amplitude of the wave and the length A-C is its wavelength (λ). The journey of the wave through A-E-B-C completing one wavelength is called one cycle. The number of cycles in one second is the frequency of the wave measured in Hertz (Hz). The velocity of light (v) is related to frequency (ν) by the relation:

$$v = \nu \lambda$$

The assumption for the study of Optical Mineralogy is that frequency, í, of a light wave is independent of the medium through which it passes. Therefore, when light enters in a mineral from air or from another mineral, it would change its velocity in order to maintain í constant when wavelength changes.

Wave Surfaces

The passage of wave motion through one complete may also be visualized as 360 rotation of a point along the circumference of a circle as shown in Fig. 1.3. However, the purpose of Fig. 1.3 is to demonstrate another important concept for us. Let the centre of the circle be considered as a source of light. Waves will emerge out from this point O in 3-D space around it and after a given time, t second, would travel to certain distance from O. Assuming that the velocity of light along all three Cartesian axes, x, y, and z are same, i.e., the medium is Isotropic, we conclude that all points in phase after t seconds will be at a distance R, and the locus of such points will be sphere. The circle in

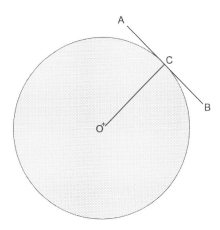

Fig. 1.3 Wave surface in an isotropic medium.

General Text Box 1.2

CHRISTIAN HUYGENS

One of the foremost scientists of the seventeenth century and a contemporary of Isaac Newton, Christian Huygens was born on April 14, 1629 in The Hague, Holland. His brilliance is obvious from a long list of concept inventions and equipment inventions. To name a few we place on record: wave theory of light, calculus of probabilities, identification of Saturn's rings and Saturn's moon, improving the power of telescope, grinding and polishing techniques of lenses, pendulum clock, checking and improving the timekeeping, etc. Louis XIV, the Emperor of France, offered him a permanent position at Paris in 1665 which he accepted. His monumental treatise, *Horologium Oscillatorium*, was published in 1673 from Paris. His contributions to telescopy were so fundamental that even today the eye pieces are known by his name: **Huygens ocular**. His *Theory of Light* was published in 1690 though deserved applause for it came posthumously. Huygens died on July 8, 1695.

in Fig. 1.3 is a section in the plane of the paper of this sphere. The surface of the sphere is termed as the *Isotropic Wave Surface*. All the waves reaching this surface are in the same phase hence, it is termed wave front, particularly, if this is the leading situation, say, after t seconds. Radius OC is coincident with a light ray that is normal to tangent ACB at C. This line ACB is called the *wave front* to the ray OC, and OC is *wave normal*. In an isotropic medium, wave normal coincides with the light ray. For a light beam there will be a large number of rays parallel to each other, hence, the wave front to the beam would be common. Many minerals, however, do not transmit light with ease in all directions. This is because transmission of light depends upon the interaction of light photon with outer electrons of an atom, and the energy of electrons varies from element to element; in addition, atoms of an element are influenced by surrounding atoms of the same or other elements. This geometric arrangement is different in each mineral (space group symmetry and interatomic bonds). We may visualize that as light travels in different directions within crystals that contain different atomic arrangements along x, y, and z directions, both intensity and velocity of light are affected. Hence, at the end of t seconds, light would not travel by a distance of R as in the case of isometric crystals, but travel distances would vary. Instead of forming a spherical wave surface, an ellipsoidal wave surface would be generated. Fig. 1.4 is an elliptical section of this wave surface in the plane of the paper. Here, again O is the source of light, OC is the path of the light ray, ACB is the tangent to the ellipsoid at C, and ON is normal to ACB. ACB is, as also in an isotropic mineral, the wave front, but now the wave normal is not coincident with the light ray. Such minerals are called *anisotropic minerals,* and the phenomenon is known as the *anisotropism.* All minerals that crystallize in isometric (cubic) system are isotropic, whereas all other minerals are anisotropic.

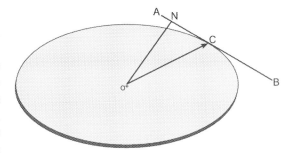

Fig. 1.4 A section of the antisotropic wave surface.

Optical Phenomena

We are concerned with some phenomena of optics that have applicability in optical mineralogy. These are *Reflection* and *Refraction* of light and their various manifestations, and *polarization* of light. You are quite familiar with these from your school curriculum but it may still be desirable to restate them.

Let us imagine that a ray of light is propagating through a given medium, say 1 (air, for example) with a given velocity, V^1; but when it reaches an *inter-phase boundary* a number of phenomena take place. Part of the light is reflected into the same medium 1 such that the angle of incidence, i, is same as the angle of reflection, r, (Fig. 1.5A). Another part of the light ray enters medium 2 but suffers a distinct change in direction depending upon how much dense both media are with reference to each other (Fig. 1.5B). The bending is controlled by *refractive indices* of the two media. The

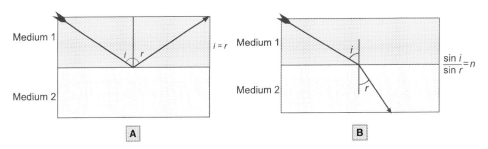

Fig. 1.5 Common optical phenomena. **A.** Reflection of light; **B.** Refraction of light.

refractive index is one of the most important optical parameter for our purpose as mineral identification is either directly done by measuring it, or by measuring other optical properties that are derived from it. It is defined as follows:

Properties of Light

$$n = \frac{V_{vac}}{V_{mat}}$$

where n is the refractive index of a mineral (or any other material), V_{vac} is the velocity of light in vacuum, and V_{mat} is the velocity through a material (mineral). Refractive index of a medium is measured by applying *Snell's Law* that relates angle of incidence and of refraction as follows:

$$n_1 \times \sin(i) = n_2 \times \sin(r)$$

where n_1 and n_2 are refractive indices of medium 1 and medium 2 respectively, and i and r are angles of incidence and refraction respectively.

If the light is travelling initially through air, *i.e.*, incident medium is air (or vacuum) then n_1 would become 1 by definition; and so

$$n_2 = \frac{\sin(i)}{\sin(r)}$$

We can easily see that refractive index bears inverse relationship to the velocity. Denser medium slows the velocity and the rarer medium increases it. When light enters from a rarer medium into a denser medium angle i > angle r, *i.e.*, light ray will bend inwards towards the normal to the phase boundary (Fig. 1.5). Just the opposite will be the case if the light ray direction is reversed. Now look at Fig. 1.6A where a ray of light is passing from a denser medium to a rarer one, and we expect it to bend away from the normal. However, in the figure the angle of incidence is such that the refracted ray has bent enough to become parallel to the phase boundary! This angle of incidence is called *critical angle*; its value depends upon the two refractive indices of the participating media. Any further increase results in further bending that simply means that light now propagates within the denser medium. It, therefore, becomes a special case of reflection and the phenomenon is called *Total (or Internal) Reflection* (Fig. 1.6B).

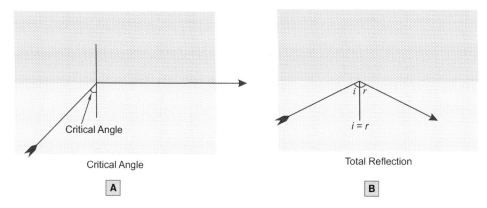

Fig. 1.6 Some important phenomena of light. **A.** Critical angle; **B.** Total reflection.

Polarization of Light

Fig. 1.7 illustrates an important optical phenomenon of interest to us. A ray of light after striking a phase boundary normally gets reflected as well as refracted. In such cases, both components are partly polarized, *i.e.*, vibrations caused by the propagation of the ray become restricted. Moreover, according to *Brewster's law*, when the angle between the reflected and refracted is 90°, the reflected ray is completely polarized, though the refracted ray is partly polarized. This is shown by superimposed dots on the reflected ray, meaning thereby that the vibrations caused are normal to the plane of the paper. For refracted ray the strokes indicate vibrations within the plane of the paper.

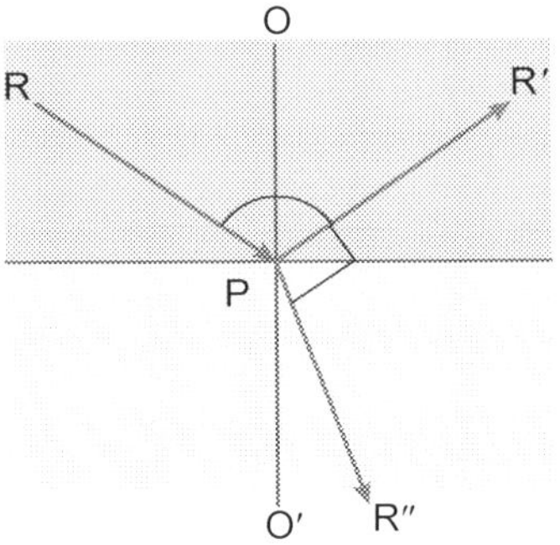

Fig. 1.7   Demonstration of Brewster's law. $\angle R'PR'' = 90°$; $\angle O'PR'' = $ Angle of refraction; $\angle RPO = $ Angle of incidence = Angle of reflection.

One of the fundamental differences between a longitudinal wave and the one of transverse nature is the relationship between the direction of oscillations and propagation. Unlike longitudinal waves, where the oscillations are restricted along the direction of propagation of rays, the transverse waves do not put such restrictions. It is because the direction of oscillation of the electric field in the case of transverse waves is not uniquely determined *a priori*. However, it is possible to restrict the oscillations in a plane perpendicular to the direction in which the waves are propagating. When this is achieved the light is said to be *polarized* and the phenomenon is called *polarization of light*. Let us imagine that the electric vector is situated in a space such that the ray is propagating along z-axis; we can resolve the electric vector along x and y as shown in Fig. 1.8. There are number of possibilities that can occur in this situation. The two components may have the same amplitude and also be in the same phase. You can see that the tip of the vector traces out a line on the x-y plane in this case; hence, this type is known as *linear polarization* (Fig. 1.8A). When the two components have exactly the same amplitude but are out of phase by $\varpi/2$ then x component may be slower or faster than y component. In either case the vector traces a circular path but for each of these two possibilities, the directions of the motion are opposite. This is known as *circular polarization* (Fig. 1.8B). When the two components are not in phase, the phase difference is not $\varpi/2$, and these two rays do not have same amplitude, the vector traces an ellipse and the phenomenon is called *elliptical polarization* (Fig. 1.8C). The interaction of the electromagnetic radiations alters direction of propagation as well polarization. Many substances are capable of decomposing the waves into two orthogonal components with different propagation effects. In such a situation a differential delay may take place; materials that transmit light in such a manner are called *birefringent*.

Properties of Light

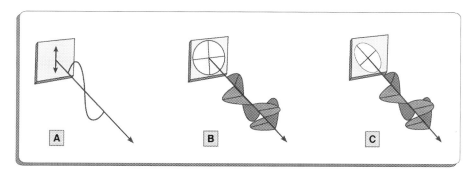

Fig. 1.8 Various forms of polarized light. **A.** Plane polarized light: The light ray passes through a filter, which has a single, preferred vibration direction; **B.** Circular Polarization: The light ray passes through a filter that has two vibration directions; **C.** Elliptical Polarization: The light ray passes through a filter which has two vibration directions.

Mineralogic studies involve the use of polarized light. Light from a natural source is commonly unpolarized, which means that the plane of vibration of E_0, the electric vector of light ray, changes its orientation very rapidly and in a completely random fashion. However, if the randomness of E_0 is checked, *i.e.*, when light becomes polarized, it is easier to understand the effects of light ray – crystal structure interaction. The influence of crystal symmetry is brought out very sharply through the propagation of a *polarized light* beam across the crystal. Hence, this phenomenon is of great importance to us. Let us imagine a device; we name it *"Polaroid"* that has the capability to dissipate E_0 vibrating in all directions except one. This direction is the direction contained within a plane called plane of polarization. Polarized beam can also be restricted to circular or elliptical modes. For optical mineralogic studies *plane polarized light* is well-suited. The restriction in the nature of vibration of light can also be brought about by processes of reflection and refraction in addition to selective absorption as mentioned above.

Double Refraction

Now, we come to a topic that you shall find interesting and engrossing, but you may not realize its great significance at this juncture. It is the phenomenon of *double refraction*. It was noted by common man and physicists that a transparent crystal of calcite (Fig. 1.9) allows two distinct well-separated images of a single object. Huygens, explanation of this phenomenon,

(Courtesy: Department of Geology Collection, DU)

Fig. 1.9 An illustration of double refraction.

widely accepted for centuries, was simple in its logic. That is, if two images are formed then there must be two rays. But there is only one incident ray; hence, calcite is capable of splitting a single ray into two rays during refraction within it. What about polarization of the refracted light rays? Yes, these are plane polarized with planes of each ray being normal to each other. The relationship between the two images is also interesting. If the calcite rhomb is rotated around an axis normal to the plane of the paper on which rhomb is placed, one of the images moves tracing a circular path with the other image remaining stationary at the centre of this circle. It was also realized that if the object was viewed through the rhomb along its *c*-axis then only one image formed that remained stationary during the rotation of the crystal. The image that remains stationary is called ordinary image, and, it can be located by following the known laws of refraction cited above. The ray responsible for this image is called the *ordinary ray*. On the other hand, the second image (the one that circles the first one) is called *extraordinary image*; *extraordinary ray* is the term for the ray responsible for this image. We can analyze this phenomenon by means of Fig. 1.10A. If we view the crystal parallel to *c*-axis then we do not see two images (Fig. 1.10B). On the other hand, the two images would lie along the line of vision when we see it normal to *c*-axis (Fig. 1.10C). Why don't these rays traversing through the same crystal at the same time interfere with each other? They do not interfere because the necessary condition for interference is that the two ray be either in the same plane or must share a component in that plane. But in the case of double refraction the rays, as postulated, vibrate in separate planes that are normal to each other.

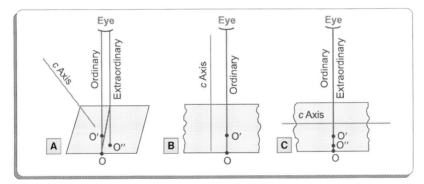

Fig. 1.10 Splitting of an incident ray into ordinary and extraordinary rays after undergoing refraction in a doubly refracting crystal. O is an object, O' is the ordinary image, O'' is the extraordinary image. **A.** The rhomb is placed with one of its {10$\bar{1}$1} faces flat on the object (random situation); **B.** Rhomb placed in such a manner that the light ray is parallel to *c*-axis (oriented situation); **C.** Rhomb placed in such a manner that the light ray propagates normal to *c*-axis (oriented situation).

Interference of Light

In the preceding section we have already noted that although there are two rays within the calcite rhomb they do not interfere. But, interference of light does take place. If there are two or more rays of light, it is obvious that energy spread by these rays will affect the medium of transmission in a resultant or combined manner. It is certainly possible to predict the response of the medium to these incoming radiations. Depending

Properties of Light

upon the properties of the two rays, like wavelength, amplitude etc., a number of situations can be visualized and analyzed. In the context of the optical mineralogy we are most concerned with a situation where the two rays are plane polarized, have the same wavelength and their planes of vibrations are coincident. This is the most common situation for which our microscopes are designed, and it is prudent that we understand interference of light in this context.

In Fig 1.11 we see three situations of two plane polarized rays propagating in the same direction and vibrating in the same plane. Fig. 1.11A refers to a case where the velocities of the two rays are same and they start from point P together. The amplitudes of the waves of the rays are shown different though it need not be so always. The dashed line is the resultant wave that has been calculated on the basis of equation given in Technical Text Box 1.2. Such waves are called waves in phase, and their displacement

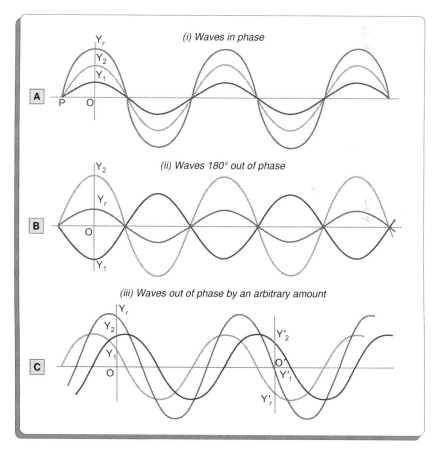

Fig. 1.11 Interference of light rays. **A.** Both rays have the same velocity of propagation with no phase difference; **B.** Both rays have the same velocity of propagation with a phase difference of 2δ; **C.** Both rays have the same velocity of propagation but one starts with a time lag with respect to the other.

Technical Text Box 1.2

INTERFERENCE OF LIGHT RAYS

Considering that wave propagation of light can be visualized as a harmonic disturbance propagating along, say, x-direction, we can write instantaneous displacement at location 'x' as:

$$\psi(x, t) = \psi_0 \sin(ùt - kx) \qquad \qquad ...(1)$$

Here, $\varnothing_0$ is amplitude of wave, $ù$ is angular frequency $= 2\delta \times í$ ($í$ is the frequency), and k is the wave vector such that $k = 2\delta/ë$

Imagine two such waves of nearly equal amplitude and same frequencies propagating along x

$$\psi_1(x, t) = \psi_{1,0} \sin\{(ùt - kx) - ö_1\} \qquad \qquad ...(2)$$
$$\psi_2(x, t) = \psi_{2,0} \sin\{(ùt - kx) - ö_2\}$$

Here, $ö_1$ and $ö_2$ are the phase angles for the two waves. The phase difference $(ö_1 - ö_2)$ remains constant all the time. The resultant displacement at a given location 'x' and time t be written as, by principle of superposition:

$$\varnothing_{res} = \varnothing_1(x, t) + \varnothing_2(x, t) \qquad \qquad ...(3)$$

or,

$$\varnothing_{res} = \psi_{1,0} \sin\{ùt - kx) - ö_1\} + \psi_{2,0} \sin\{ùt - kx) - ö_2\}$$
$$= \varnothing_{1,0}(\sin ùt \times \cos ö_1 - \cos ùt \times \sin ö_1) + \varnothing_{2,0}(\sin ùt \times \cos ö_2 - \cos ùt \times \sin ö_2)$$
$$= (\varnothing_{10} \cos ö_1 + \varnothing_{20} \cos ö_2)\sin ùt - (\varnothing_{10} \sin ö_1 + \varnothing_{20} \sin ö_2)\cos ùt$$

Let $(\varnothing_{10} \cos ö_1 + \varnothing_{20} \cos ö_2) = \varnothing \cos ö$

and $(\varnothing_{10} \sin ö_1 + \varnothing_{20} \sin ö_2) = \varnothing \sin ö$

Therefore, the resultant displacement at location 'x' and instant 't' can be written as:

$$\varnothing_{res} = \varnothing \cos ö \sin ùt + \varnothing \sin ö \cos ùt,$$

or

$$= \varnothing \sin(ùt + ö) \qquad \qquad ...(4)$$

It is obvious from equation 4 that a superpositioning of original two waves results in the formation of a new single wave. The new wave is harmonic with the same angular frequency, $ù$, as the original two waves but with a different amplitude, $\varnothing$, and phase '$ö$'. The situation is repeated at another location separated by $ë$ distance.

The new amplitude is given by:

$$\varnothing = [\varnothing_{10} \wedge 2 + \varnothing_{20} \wedge 2 + 2\varnothing_{10} \times \varnothing_{20} \cos(ö_1 - ö_2)];$$

$\varnothing$ is maximum when:

$$\cos(ö_1 - ö_2) = 1;$$

or,

$(ö_1 - ö_2) = \delta$ (even integer); *i.e.*, the two original waves are in same phase at location 'x. $\varnothing_{max} = \varnothing_1 + \varnothing_2$, *i.e.*, it is equal to sum of two original amplitudes.

$\varnothing$ is minimum when:

$$\cos(ö_1 - ö_2) = -1$$

or

$(ö_1 - ö_2) = \delta$ (odd integer); *i.e.*, two original waves are in opposite phase at location 'x'.

$\varnothing_{min} = \varnothing_1 - \varnothing_2$, *i.e.*, it is equal to difference of two original amplitudes.

The above exercise explains that the interference pattern will be a series of bright bands separated by dark bands. The two consecutive bright or dark bands are separated by $ë$ distance.

(amplitude), y values, in the equation are added to get the final displacement. This is an algebraic addition where displacement as a vector will have positive sign if it is above the ray line, but negative if it is below the ray line. Hence, for a case like that of Fig. 1.11A the intensity, being governed by the amplitude, would get enhanced. However, in Fig. 1.11B the resultant amplitude is zero as crest of one wave is matching with the trough of the other. Such waves are called as differing with each other by a phase difference of 180° or 2δ radians. When the amplitudes are equal but opposite, the two waves will cancel each other with no light emerging out. Fig. 1.11C shows the two rays starting at different times along the same ray direction. The calculations for the resultant wave still follow the equations given in Technical Text Box 1.2. The phase difference between the two rays will be intermediate to two extreme values of 0 and 2δ radius.

We may now describe a situation where the two rays may not have the same wave length. The rules of calculating the resultant wave are still the same as given above. We all know that a wavelength of light ray produces sensation of color, whereas amplitude gives a sensation of intensity. Hence, in this general situation, two differently colored rays will produce a ray of third color of different intensity.

The polarizing microscope is designed in such a manner so that a user can easily create the above-mentioned situations to learn the response of minerals through which the rays transmit. You must have guessed that behaviour of light through different minerals will be different. The description of this behaviour is known as the Optical Properties of minerals.

Summary

A review of simples laws of optics learnt at school level is presented. Optical mineralogy is the study that combines interaction of light with a given mineral and its internal structure. The most important optical property of a mineral is its refractive index. Its value depends upon the crystalline nature of minerals; at the same time for many minerals, the value of refractive index depends upon the ray direction with reference to crystallographic directions. Minerals are classified as isotropic (with constant refractive index) and anisotropic (where refractive index varies). Anisotropic minerals polarize light into planes known as vibration planes. These minerals show double refraction, *i.e.*, formation of two images. Another important phenomenon of interest is that of interference of light that involves combining two polarized light rays into one. The amplitude of the resultant ray depends on the phase difference between the initial two rays.

Important Terms

Amplitude of the wave 7
Anisotropic minerals 10
Brewester's law 12
Coherent scattering of the radiation 7
Critical angle 11
Double refraction 13
Extraordinary ray 14
Frequency 8
Hertz 8
Incoherent radiation 7
Inter-phase boundary 10
Interference of light 14

Isotropic wave surface 9
Longitudinal waves 7
Ordinary ray 14
Phase difference 12
Reflection 10
Refraction 10
Refractive index 11
Total reflection 11
Transverse waves 7
Wavelength 8
Wave surfaces 8

Questions

1. What is scattering of light by an atom?
2. How many types of scattering of light by an atom are commonly observed?
3. Describe various components of a transverse wave.
4. Draw and describe electromagnetic spectrum.
5. What is meant by the term isotropic wave surface?
6. Is propagation of light affected by the internal structure of the medium?
7. Describe the phenomenon of anisotropism.
8. Derive Snell's Law of refraction.
9. How does total reflection achieved?
10. Define the term polarization of light.
11. What is Brewester's Law?
12. Explain double refraction.
13. Diagrammatically show ordinary and extraordinary rays.
14. Describe the phenomenon of interference of light.
15. If a beam of light consists of two or more rays with different wavelengths, then how does the interference of these affect the final character of the beam?

Suggestion for Further Reading

Beginning Physics II: Waves, Electromagnetism, Optics and Modern Physics, 2007. Alvin Halprin, Schaum's Outline, (Paperback).

Optics, 2007, Eugene Hecht., Schaum's Outline, (Paperback).

2
Sample Preparation for Transmitted Light Microscopy

Learning Objectives

- The purpose of thin section preparation for optical mineralogy
- Techniques of thin section preparation
- Problems presented by individual samples and their elimination

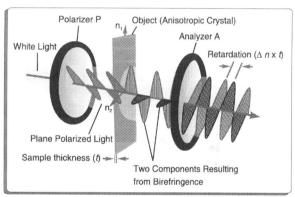

A birefringent crystal between two polars.

(Courtesy: Leitz Co.)

Prologue

A transparent substance allows rays of light to transmit through it so that objects situated behind or beyond it can be distinctly seen. Our purpose is not to see 'behind or beyond' but within the substance. Nevertheless, quality of transparency is necessary and is not obvious in samples picked up from outcrops. Fortunately, most of the rock-forming minerals and rocks can be processed in such way so that light is transmitted through them. But doing so is a matter of skill, as well as that of scientific knowledge. The quality of a sample prepared, affects the quality of microscopic examination of rocks and minerals.

Chapter's Outline

- The polarizing microscope – primary equipment for optical mineralogy
- The purpose of sample preparations
- Nature of samples
- Size of samples
- Thickness of a sample and the concept of transparency
- Polishing and mounting rock chips

© The Author(s) 2023
P. K. Verma, *Optical Mineralogy*,
https://doi.org/10.1007/978-3-031-40765-9_2

Introduction

The subject-matter of the present book is primarily concerned with petrological microscope, or simply polarizing microscope (*see* Chapter 5), and, therefore, the sample preparation described in this chapter is with reference to studies conducted with the help of a polarizing microscope. It is important that beginners in mineralogy-petrology should acquaint themselves with these techniques as the quality of observation is directly related to the quality of the given thin section. Although the steps are more or less similar in various sample preparation techniques, yet, the variation must be adopted to suit individual physical property of a given rock sample. The influence of physical properties of rocks, especially, tenacity, can never be overestimated; hard rock is difficult to thin but a soft rock crumbles easily, and, so is washed out even before polishing begins; so each new rock specimen presents its own peculiar problems, hence the need to innovate is always there.

The Polarizing Microscope – Primary Equipment for Optical Mineralogy

Three kinds of microscopes: *binocular microscope, petrological or transmitted light microscope,* and *ore microscope or reflected light microscope*, are commonly employed in optical mineralogy. By binocular microscope is meant a microscope which uses non-polarized beam of light, and is equipped with stereo-zoom optics for a clear 3-D vision of small grains, rock fragments, or even small samples. There are two oculars for convenience of observation, hence, the name given binocular microscope. The petrological (transmitted light) microscope may also have two (in many models even three) oculars, but has several other provisions for a number of observations of importance to mineralogy and petrology. All the observations are taken when the light beam, in general, is plane polarized, and, is transmitted through the mineral. On the other hand, if the light rays are reflected from the surface of the mineral, and, then viewed through an ocular or binocular, the microscope is referred to as the ore microscope or reflected light microscope. Many present day models of microscopes combine reflected light and transmitted light arrangement; this has several advantages; logistically it saves lab space, and is cheaper than buying two separate microscopes. However, in the experience of the author separate models for each observation are better, as flipping between transmitted and reflected light with different samples may require frequent set-up changes causing strain on the long life of the equipment.

All three kinds of microscopes are necessary in a mineralogy laboratory. The binocular microscope is useful in studying individual grains in an aggregate, to handpick desired grains for further work, such as geochronological determinations, and to carry out staining experiments. It is also useful in mounting grains on spindle stages etc. The reflected light microscope is useful in studying what we call opaque minerals. On the other hand, mineral identification, beyond hand specimen scale is routinely done by

Sample Preparation for Transmitted Light Microscopy 21

means of polarizing microscope whose utility extends much further into petrology as you will realize by going through this book. Therefore, this chapter is primarily directed towards sample preparation for the polarizing microscope.

The Purpose of Sample Preparations

A chunk of mineral or rock broken away from a ledge cannot be directly studied by means of a polarizing microscope. There are two obvious constraints in our study; the first is size of the sample stage of the microscope, and the second is the condition of transparency for the light to travel through the sample. After the study is completed, there is a need to properly store the sample in a well-labelled location. This is another aspect which is not very obvious, *i.e.*, the convenience of handling the sample, and storing and cataloguing for future retrieval. We shall learn in later chapters that some optical properties of minerals are dependent upon sample preparation. Hence, there is a need for us to follow a set procedure for sample preparation so that our observations are compared and evaluated without confusion. A caution bell may be sounded here and be taken seriously that the preparation of samples for microscopic studies is a partly destructive method in the sense that a good part of the sample is always lost during the preparation. Careful attention would reduce the loss but cannot entirely eliminate it[1].

There are three steps involved in the sample preparation. The first is to cut the sample to right size comparable to a common microscope stage, then to make it transparent, and finally, to preserve it in a convenient form.

Nature of Samples

A variety of material can be viewed under polarizing microscope. The most common is a thin section of a rock, but thin sections can also be prepared from other materials as well. For example, geologists working on the Quaternary sediments often have loose soil cover to work with. A binocular stereo-zoom microscope is often used for such studies, yet one may find a thin section study of soils or sand very useful. Sedimentary petrologists disintegrate sandstones and study thin sections prepared out of various sieve fractions including heavy mineral assemblages. We may wish to make a thin section of one or more grains of a given mineral, or, synthetically prepared crystals of a compound. Thin sections of fossils may also be made for studying the internal structure of shells, plant fossils, fossil bones etc.

Routinely, we shall be concerned with studying chips of rocks or minerals, but these also have a wide range in physical properties that govern the easiness of sample preparation. For example, homogeneity of grain size may be present; alternatively, great variation in grain size is present. Homogeneity of structure is another consideration.

---

1. It is crucial when microscopic examination is undertaken of moon rocks, meteorites, or rocks that are collected from wilderness region such as the Antarctica; in other words, rocks that cannot be collected again easily.

A rock sample may contain large number of closely-spaced cleavages or schistosity planes; a sample may be banded with individual bands ranging from a millimeter to over a centimeter in thickness. These may present challenges because of contrasting tenacity and hardness of minerals.

Size of Samples

Fig. 2.1 shows two examples of different sizes broken away from the rock ledges (outcrops) that are brought for examination. As you would notice the sample of Fig. 2.1A is rather large, whereas that shown in Fig. 2.1B is a small chip; most samples would be a bit larger than that shown in Fig. 2.1B. We definitely cannot have a sample that is larger than the microscope stage for microscopic observations. Furthermore, often, the microscopic examination leads us to obtain data by means of electron probe, Raman probe, heating-freezing experiments, atomic force microscopy or image analytical work. These equipments may have constraints on the size of their sample chambers, so it is

(Courtesy: K.K. Singh, Geology Department Collection, DU)   (Courtesy: K.K. Singh, Geology Department Collection, DU)

Fig. 2.1 Two common sample sizes brought to the lapidary. Sample size obtained from an outcrop and brought in for making thin sections depends upon several factors including the objectives of the investigation. **A.** A large sample of nearly 30 cms across; **B.** A rather small sample of nearly 10 cms across.

always advisable to keep sample size within 1cm × 2cm size. Now, this size is awfully small compared to the samples shown in Fig. 2.1 from which we have to make a choice to chip off a small part. Therefore, we must choose a portion that is representative; otherwise someone else might be clever enough to make discoveries from our collection at a later date by making an alternative choice of chip to study! Hence, a close physical examination in the lab of cleaned and washed sample is necessary before deciding the portion from which the chip is to be drawn. A comparison with field diary helps as experience shows that ideas springing up during the field work and carefully noted could turn out to be the best. The purpose of microscopic examination also guides the location of the chip. For example, a routine petrographic examination allows a lot of freedom; for such examination, it is often advisable to make more than one chip thereby

Sample Preparation for Transmitted Light Microscopy 23

reducing the risk of missing out something important. When one is interested in a specific mineralogic or petrologic aspect, such as looking for inclusions within porphyroblasts, or phenocrysts (large size grains), then we are bound to prepare thin sections from where these occur most. At times, we would like to investigate the interphase of two bands in a rock, apices of crenulations, pressure shadows etc. Likewise, we may be interested in oriented sections of minerals and rocks, *i.e.*, a section cut in a specific manner, for example normal to *z*-axis of a mineral or normal to foliation planes. We must decide these options beforehand from the efficiency point of view.

A good thin section preparation lapidary has a set of cutting machines though two may be suffice if the workload is less. A large cutting wheel (30cm radius) shown in Fig. 2.2A is good enough to cut large pieces; though its installation is rather cumbersome as a solid raised platform is constructed on which it is bolted. A three-phase connection of electric mains with a standard MCB is necessary since a heavy duty motor (one horse power) rotates the cutting wheel. The working area (Fig. 2.2B) should be fairly large and well-ventilated since dust and rock splinters fly in truly 3-D space! A continuous

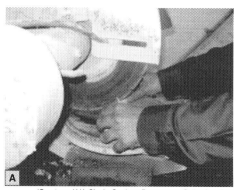
(Courtesy: K.K. Singh, Geology Department Collection, DU)

(Courtesy: K.K. Singh, Geology Department Collection, DU)

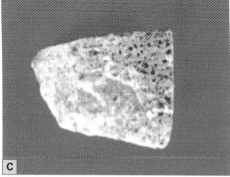
(Courtesy: K.K. Singh, Geology Department Collection, DU)

(Courtesy: K.K. Singh, Geology Department Collection, DU)

Fig. 2.2 A. The large cutter in action; **B.** An overall view of the 30 cm cutter and its location; **C.** A small sample obtained from the large cutter; **D.** The 10cm cutter as a desktop.

water supply and disposal drainage with trappings for rock outwash should complete the system.

It is suggested that the cut pieces from the large cutter be preserved and individually labelled for future use. One such rock piece (Fig. 2.2C) now may be mounted on a small cutter that has 10cm radius cutting wheel (Fig. 2.2D). You will notice that it is almost a desktop model with electric and water supply connections close by. The piece obtained from the large cutter, washed, now may be clamped on its stand, water jet be started, and then the motor be switched on. When a small sample is supplied then we may not require the large cutter, and the sample be directly clamped in this cutter for getting a chip of about 3cm × 3cm cut area. It may not be exactly rectangular though it would help if it sort of resembles a rectangle!

Thickness of a Sample and the Concept of Transparency

Once a choice of a chip is made then we must turn our attention to how thick the chip should be? A related question is about the extent and level of uniformity of thickness is to be obtained? Thickness of the chip is important because that determines the length of the path of light through our sample. We have already discussed in the previous chapter that light travels in a medium through a series of interactions with electrons of the medium. More the light travels through the object more it interacts with the electrons. These are the electrons in the outer shells of atoms of the elements held in the crystal lattices. Each interaction transfers part of the energy of light to electrons, and a small part of that is converted into other forms of energy which is not retrievable as light. So the light suffers loss of intensity while traversing through a medium. The loss may be formidable and complete so that no through transmission results. The sample is termed *opaque* to light and is not suitable for transmission light microscopy. Long back it was observed that a thickness of 30μm (*i.e.* 0.03mm) is good enough to allow overwhelming number of minerals to transmit sunlight (daylight). Hence, by convention chips are grounded to this thickness; from practical consideration, it is easily achievable as we shall describe in the following pages. All optical properties of minerals listed in this or any other book are based on this thickness. In case your observations do not match with those described in the book, it would help if you check the thickness of your thin section. Since, thickness is specified, it is very important that the chip should show uniform thickness of 30μm. The accuracy should be about 1μm level or better.

Let us assume that the by the use of both cutters described earlier we have a rectangular chip which has a thickness of about 1/2 cm; if the thickness is more than use the small cutter once more to get this thickness (Fig. 2.3). Now in order to finally reduce the chip to 30μm thickness the first step would be polishing described in the following section.

(Courtesy: K.K. Singh, Geology Department Collection, DU)

Fig. 2.3 The approximate thickness of the rock chip before mounting on a glass plate.

General Text Box 2.1

HEINRICH FERDINAND ROSENBUSCH

Prof. Heinrich Ferdinand Rosenbusch was born on June 24, 1836 in Einbeck, district Northeim, in southern Lower Saxony, Germany. The town of Einbeck is famous as the birthplace of Henry Muhlenberg who established the Lutheran Church in the United States of America way back in the eighteenth century. Einbeck is also known for one of the oldest beer brewary in Germany. Young Heinrich had his early education in this town. For his higher education, he joined Heidleberg University.

He distinguished himself as a petrologist and his fame as a teacher and researcher spread far and wide. His keen sense of learning newer aspects of science was hallmark of his carrier. He was perhaps the first petrologist to recognize the importance of Nicol prism for obtaining polarized light. He perfected the compound microscope with Nicol prism, and began the systematic optical examination of minerals. This resulted in his monumental work, *Mikroskopische Physiographie der Mineralien und Gesteine*, 4 Vols., 1873-1877.

Heinrich Rosenbusch was invited to serve on the teaching faculty of Heidleberg University in 1877. A number of brilliant scholars came out from his research group. He is considered the father of modern petrography. He served Heidleberg all his active life retiring in 1908. He breathed his last on January 20, 1914 at Heidleberg.

Polishing and Mounting Rock Chips

The chip of right size obtained as outlined above is now to be polished on its larger surface. The reasons for the polish are to be dealt with soon, but before the step for polishing is undertaken, we must first examine the tenacity of the chip, *i.e.*, whether it would polish well or just wither away during the polishing. If the rock is hard and compact then we may proceed right away. The soft rocks, highly cleaved samples, rocks composed of minerals of contrasting hardness (for example, fine grained muscovite with porphyroblasts of garnet or staurolite embedded) present enormous problems requiring great expertise and great patience. A series of experiments to improve the polishing quality may have to be undertaken though first step, commonly, is *cooking* the chip in hardening medium for some time. A common medium is *resin*, which costs very little though fumes are a bit messy, and the cooking is done within a *special fuming cupboard*. Several hours of cooking time is recommended. The procedure is simple as a metal can (often a used container) is put on the burner with chunks of dry resin along with several rock chips. It takes anywhere between 3 to 10 hours of heating (boiling). After the cooled chip is taken out on a grinder, it would be obvious in a short while whether the sample would take up polish or requires another cooking session. Don't loose your patience if the sample needed three or more sessions of cooking.

A motorized grinding wheel, commercially available, is used first to scrap the dried resin, and then remove angularities and grooves (arising out of cutting wheel) from the chip; normally you hold the chip in your hand taking precaution about not hurting your fingers. Initially water dripping from a trap or simple periodically poured by you on to the chip is enough; but with the progress of grinding, a grinding powder made out of silicon carbide of mesh size 120, and later on with powder of mesh size 220 sprinkled on the grinding wheel along with drops of water are employed. Slowly and in 2 to 15 minutes depending upon the hardness of the rock, the surface begins to acquire a polish. Frequent washing of the chip and testing for the quality of polish that the sample is progressively acquiring, will save avoidable effort; a good polish is indicated by the chip surface when it reflects light after the chip is washed and dried. Further, fine polishing is required, which can be accomplished by hand (Figs. 2.4A and B) on an iron plate with silicon carbide powder of 400 mesh size, transferring it later to a glass plate

(Courtesy: K.K. Singh, Geology Department Collection, DU)   (Courtesy: K.K. Singh, Geology Department Collection, DU)

Fig. 2.4 A. A thick glass plate to be used for fine polishing of the chip; **B.** A closer view of the fine polishing in action.

with powder of 600 and 800 sizes after reasonable polish. The manual procedure is to use a thick glass sheet, about 3-5mm thick roughly a square of foot in size. A small spatula is used to pour the polishing powder on the glass sheet and moisten with a few drop of water. The polished surface of the chip is put on this moist heap and with slight pressure of fingers the chip is polished with periodic check to note the progress. Some of these steps may be accomplished by means of an automatic machine. The automatic machine has clamps for holding samples and a motor motion allows the sample to be rubbed on a hard surface with the help of polishing powder (Fig. 2.5). For a perfect planar surface with a good reflecting capability, the chip is made to go through another

Sample Preparation for Transmitted Light Microscopy

Technical Text Box 2.1

MESH SIZES

It is a common knowledge that a powder is characterized by its chemical composition and by the size of its particles. It must be clearly understood that there is always a range of size of particle in any given sample of powder. Hence, the particle size of a given powder is actually the average grain size of the sample particles. From the lapidary point of view, which is our aim, powders are used as abrasives, and also as polishing medium. It is very essential that the choice of a particle size should be right for a given surface of a given rock. Too coarse material would damage the rock chip and too fine powder would be just a waste of material and time.

There are a number of mesh sizes that have come to be known as standardized mesh sizes, *e.g.*, BS, ASTM E 11-70, ISO 3310 etc. The designation of a particular size by a given number depends upon whether that number refers to number of wires per linear inch or the number of opening per linear inch. It is a common practice to categorize a given sieve by its mesh size, and also print the opening dimension as well. For example, a sieve of Tyler 80 is that with an opening of 0.178mm.

There is another pattern of reference for the particle size that mineralogists and petrologists often follow while crushing a rock for various examinations. Routine refractive index determination is best carried on grains that are – 80+120, meaning thereby particles that would pass through a sieve of size 80 but would be retained by sieve of size 120; this would correspond to openings of 0.125 – 0.178mm. For powder XRD, mesh size should be >200, *i.e.*, <075mm. The following table lists certain sizes and equivalents.

Sieve size (mm)	BSS (approx.)	Tyler (approx.)	US
2.00	8	9	10
0.853	18	20	20
0.599	25	28	30
0.500	30	32	35
0.422	36	35	40
0.251	60	60	60
0.178	85	80	80
0.152	100	100	100
0.125	120	115	120
0.075	200	200	200
0.044	350	325	325
0.037	440	400	400

cycle of grinding, first with emery paper of suitable efficiency, then a variety of diamond paste are utilized. Finally, alumina powder of one micron grain size made moist with few drops of water is used. An excellent reflection with smooth surface, (*i.e.*, no pit marks etc.) means that the first step towards an excellent sample has been taken.

Now the chip is ready for *mounting*, *i.e.*, to be placed and glued to a glass plate. As mentioned above, the size of the sample is appropriate with reference to equipments to be used in the detailed

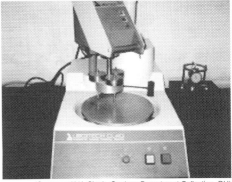

(Courtesy: K.K. Singh, Geology Department Collection, DU)

Fig. 2.5 An automated multi-sample grinding wheel.

examination; the same rule is applicable to the size of the glass plate. The thickness of the glass plate should be 2 to 3mm as most lens systems in microscopes are designed with this thickness in mind. Normally, these are available commercially, because of extensive use for biological microscope examinations in schools, colleges, hospitals etc. Any adhesive material that is transparent, colorless, isotropic, non-reactive to the chip, and can withstand high intensity, laser and electron beam is suitable. Currently, the most popular adhesive is *araldite*, which is available in all neighbourhood hardware shops. Apply the adhesive as a smooth layer on the chip, and, then with polished surface down on the glass plate the chip is put on the plate, and pressed with fingers to make sure that no air bubble is trapped. Presence of air bubbles is not only annoying to the user but also decreases the strength of the glue. Some companies supply presses where between the two disks the mounted slides are placed. A set of springs gently screwed ensure that a slight pressure is maintained.

Pressed mounted chips are left overnight to ensure its firm association with the plate. Then the steps outlined above for grinding and polishing are repeated till a smooth surfaces is generated opposite to, but parallel to the glued surface. Here, again, motorized polishing is becoming common. Many of these machines have a provision (Fig. 2.5) for screw gauge which is set for 0.03mm thickness. The polishing progresses till the desired thickness is achieved. This is important as all optical properties of minerals described in books are based on studies of samples of this thickness.

In order to avoid dust etc. getting on the surface of the mounted section, a thin glass sheet called *cover slip* (~0.02mm) is glued on the chip; the size of the cover slip should be more than that of the rock chip under investigation. All microscope lens systems have a provision for this extra path introduced by the cover slip. However, now a day, modern day microprobe techniques such as LA-ICP-MAS, EPMA, SHRIMP, Raman probe and AFM investigations are done on the chip surface, so the cover slip is not put. If it is known that the thin section is being prepared for such a purpose, then the top surface of the mounted chip is subjected to further polishing. This is done with special pastes, cloths and special leather akin to that used in preparing polished sections employing dedicated polishing machines (Fig. 2.6).

When the above mentioned polished is not done, light gets scattered from the grain edges so the view is not exactly similar to the case when the cover slip is put. Normally, such examinations are carried out after wetting the surface of the chip by a small drop of water, which mimics cover slip. It should be wiped out after the use so that its vapours do not cause damage to objective lens. Sample number and other information may be engraved on a corner of the mounting glass plate or on a sticker and then glued to the sheet.

The sample is now ready for microscopic examination.

(Courtesy: K.K. Singh, Geology Department Collection, DU)

Fig. 2.6 A polishing machine for preparing extra polished thin sections.

Summary

Procedures for sample preparation for studying optical properties under polarizing microscope are described in this chapter. The mineral or rock sample has to be reduced to a size comparable to the sample stage of microscope. The second step is to make the sample transparent since the optical properties are based on transmission of light. There is also a need to enclose the sample so that it is properly stored. Thin sections can be prepared out of loose sand, soils, fossils and also of rocks. Some rocks may contain one mineral in predominance or it may consist of large number of minerals. In the second case it could be a difficult job to make thin sections because of contrasting physical properties. In order to make a thin section, a chip of a given rock is cut by using appropriate cutters such that the surface exposed by cut is somewhere between 3 to 5 cm^2. This surface is well-polished by means of a number of grinding powders of different grain sizes, then, mounted by means of appropriate glue to a transparent glass slide. Then manual or automatic grinding is carried out till 30 μm thickness is achieved. A cover slip consisting of transparent glass of 0.02mm thickness is glued on top of the chip.

Important Terms

Araldite 28
Binocular microscope 20
Cooking resin 25
Cover slip 28
Diamond paste 28
Grinding powder 26
Mounting rock chips 25
Opaque minerals 20

Ore microscope 20
Petrological microscope 20
Polishing machine 29
Polished sections 29
Reflected light microscope 20
Special fuming cupboard 25
Transmitted light microscope 20

Questions

1. Why we can't observe a sample directly under microscope in the form in which it was obtained from an outcrop?
2. What are the three basic steps involved in the sample preparation for optical studies?
3. Write down factors that would govern the size selection of a sample for microscopic studies.
4. What is the thickness of a rock chip in thin sections? What were the reasons for choosing this thickness?
5. Why should physical properties of rocks influence the technique of thin section preparation?
6. Briefly describe the procedure for mounting a rock chip on the glass plate.

Suggestion for Further Reading

Hutchinsons, C.S., 1972, Petrographic Techniques, John Wiley.

Johannsen, A., 1918, Manual of Petrography Methods, McGraw-Hills.

3
Refractometry

Learning Objectives

- Importance of refractive index in controlling optical properties of minerals
- Qualitative methods for estimating refractive indices of a mineral
- Quantitative methods for determining refractive indices of an isotropic mineral

Prologue

The satellite, fully 150 kilometers in diameter, was close enough to Euterpe to shine like a brilliant dot of sparkling light. It was so close to the planet that one could see it sweep west to east across the sky, outstripping the planet's slow rotational motion. It brightened as it rose towards zenith, and dimmed as it dropped towards the horizon again. One watched it with fascination the first night; with less the second; and with a vague discontent the third – assuming the sky was clear on those nights, which it usually wasn't.

— Isaac Asinov in Robots and Empire

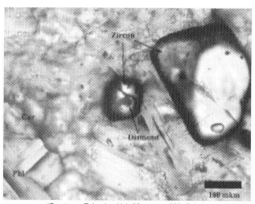

(Courtesy: *Episodes*. Vol. 26, pages 292. Sobolev et al. 2003)

A photomicrograph of a marble from Kazhakistan illustrating the refractive index-relief relationship. The matrix is of carbonate (dolomite) and phlogopite. A grain of zircon occurs as an inclusion within the carbonate. This grain also contains an inclusion of microdiamond. The inclusion zircon appears raised (high relief) and the diamond inclusion exhibits even higher relief.

Chapter's Outline

- Relief
- Becke line
- Oblique illumination method
- Refractometers

© The Author(s) 2023
P. K. Verma, *Optical Mineralogy*,
https://doi.org/10.1007/978-3-031-40765-9_3

Introduction

We have already stressed in the first chapter the importance of refractive index of a mineral in optical mineralogy. You also know about its application in forming images from lenses. In this chapter, we deal with some important applications of refractive index that will actually help us in identifying minerals. Your attention is drawn to the Appendix N in which important rock forming minerals are listed in order of increasing refractive indices. It is advised that you go through this alone and with friends once a day regularly till the order of minerals is known to you by heart. We shall first see the manifestation of refractive index of a given mineral under microscope that should allow us to make a qualitative assessment of the index value; in several cases this would suffice. Determinative methods would follow.

Relief

An important optical property of a mineral is its relief that arises because of its refractive index. A simple experiment may come in handy to start with for explanation of relief. Take a beaker and partly fill it with water; put a cover slip glass in it. Then look down in the beaker by holding it higher than the eye level. You shall see what you have done, *i.e.*, a cover slip at the bottom of the beaker (Fig. 3.1A). Now repeat this experiment with another beaker filled with, not water, but carbon tetrachloride (Fig. 3.1B). Where is the cover slip now? What happened to it? No! It cannot just disappear; it goes hiding

 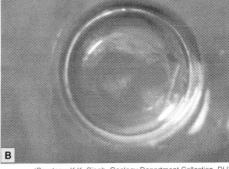

(Courtesy: K.K. Singh, Geology Department Collection, DU) (Courtesy: K.K. Singh, Geology Department Collection, DU)

Fig. 3.1 A. A beaker filled with water with a glass slide in the bottom; **B.** The same beaker now filled with carbon tetrachloride and the same glass slide with practically no outline visible.

because its refractive index being 1.52 is almost same as that of carbon tetrachloride. The lesson to be learnt from such an experiment is that the light beam uses refractive index to mark the boundary separating two transparent substances (or phases). The

Refractometry

boundary as it appears under microscope is called the *relief*[1]. Both glass slip and carbon tetrachloride having same refractive index show no boundary or relief against each other. But with water ($n = 1.33$) and glass slip there would be a strong relief, and the cover slip shall be prominently seen. Figure 3.2 shows grains of a number of minerals embedded in epoxy which has refractive index of 1.51. Do you notice any difference between these examples? The first point one may note is the shape of different grains; some amongst you may also note differences in size. But, a careful look will tell you that these grains also differ from each other in the sharpness of their grain boundary that is each of the six minerals shown differ from each other in their relief. The minerals are arranged in decreasing order of their relief. Garnet has very high relief (Fig. 3.2A), kyanite has high relief, tourmaline moderate. Quartz has practically no relief (the case is similar to glass slide in carbon tetrachloride). Then we have cases of talc and fluorite (Figs. 3.2F and G) which have negative relief. The implication of Fig. 3.2 is that a grain can be, easily identified by its relief. In practice, this may not be so simple since human eye has limitations for recognizing slight differences in relief of two minerals. Yet, with a little practice that will come from your spending reasonable time with the microscopic examination regularly, you shall develop the sense of differentiating between 'reliefs' of different minerals. In your own mind you must categorize minerals with *high* or *strong relief, moderate relief,* and *low* or *faint relief.* The term *negative relief* is used when the refractive index of the mineral is different from that of the epoxy but is lower than it. Similarly, we shall speak about *positive relief* when the mineral has a higher refractive index than the epoxy.

The suggested simple experiment with a beaker and cover slip glass has a very useful practical application. It means that relief of a mineral can be made to disappear if it is immersed in a liquid whose refractive index is same as that of the mineral. This is the principle of *liquid Immersion Method,* which is dealt with in a subsequent section.

Becke Line

In view of the importance of relief of a mineral we are tempted to explore the optical phenomenon in the neighbourhood of a grain boundary, particularly because it is our intention to utilize the relief for quantitative determination of refractive index. Let us take a thin section of a rock where two grains of two different minerals are in contact with each other (Fig. 3.3). Further, the mineral M_1 has a refractive index of 1.51($\approx$ refractive index of epoxy), and the mineral M_2 has refractive index of kyanite (≈ 1.724)

---

1. The degree to which mineral grains stand out from the mounting medium is called relief (Nesse, 1985).

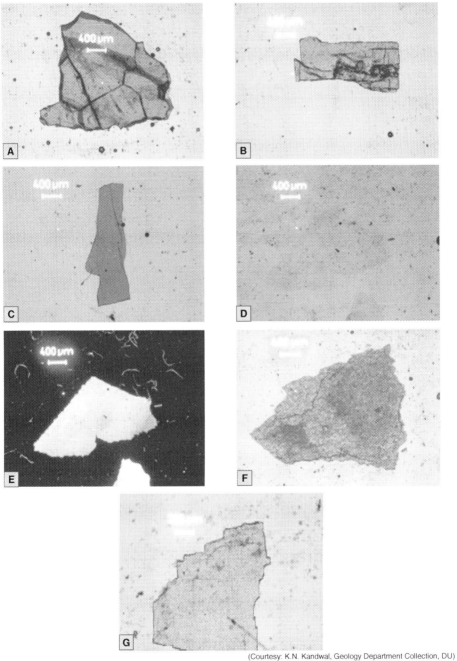

(Courtesy: K.N. Kandwal, Geology Department Collection, DU)

Fig. 3.2 Some representative examples of relief of minerals. **A.** Almandine garnet (~ 1.830) very high; **B.** Kyanite (~ 1.724) high relief; **C.** Tourmaline (~ 1.632) moderate relief; **D.** Quartz (~ 1.554) under plane polarized light, practically no relief; **E.** Quartz (same grain) in cross polars; **F.** Talc (~ 1.539) low negative relief; **G.** Fluorite (~ 1.433) high negative relief.

Refractometry

as in Fig. 3.2B. Normally, we ignore the cross-section of thin sections and deal with it as if it is a plane parallel to the plane of the glass plate on which it is glued; however, here, we shall consider the cross-section and the passage of light through it as shown in Fig. 3.3. ABCD is the cross-section of the thin section consisting of M_1 and M_2. X-Y is the grain boundary along which we shall investigate the propagation of a light beam. For the sake of simplicity let us, also, assume that both the minerals are isotropic so that we are spared of considering changes in the value of refractive indices in different directions. Let us further assume that refractive index of M_1 is nM_1 and that of M_2 is nM_2; furthermore let $nM_2 > nM_1$.

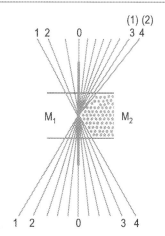

Fig. 3.3 Explanation of Becke Line formation. The phase boundary between M_1 and M_2 is assumed vertical.

If the beam of light is conical, then different rays will make different angles of incidence as shown in Fig. 3.3 where rays are marked as 1, 2, 3, and 4. Rays 1 and 2 enter the thin section through M_1 and 3 and 4 enter through M_2. The rays will be refracted in accordance with Snell's Law. Once inside the respective minerals the rays would ultimately intersect the grain boundary, which let us say is vertical as shown. Rays 1 and 2 would undergo part reflection at the surface and part refraction. The rays that enter M_2 would also exhibit similar phenomena except those like 3 and 4 that strike the grain boundary at an angle greater than the critical angle for M_2, which in this case is ~60°. These will suffer total reflection, and coalesce with the refracted components of rays from M_1. As a consequence, a bright band will appear somewhat close to the top of the thin section, though not coincident with it[2]. The band can be brought into focus by raising the objective. This bright band is called the *Becke Line,* and as shown on Fig. 3.3, it is always formed within the grain with the higher refractive index, but is situated in the upper part of the section.

Another contribution for the formation of Becke line comes from the optical character known as *'lens effect'*. The effect is generated by the fact that most minerals do not have uniform thickness along any given direction; the grains are thicker in the middle than at the edges. In other words, the grains approximate like a double convex lens as shown in Fig. 3.4. With individual grains immersed in oil this phenomenon is very conspicuous, although it is also observed in thin sections as well. The nature of refraction by these grains is either converging or diverging depending upon the relative refractive

---

2. The refracted rays always form 'raised' images.

General Text Box 3.1

FRIEDRICH JOHANN KARL BECKE

Friedrich Johann Karl Becke was born on December 31, 1855 in Prague, which was a part of the Austro-Hungarian Empire. After early education in Prague, Becke moved to Vienna. A very keen naturalist, his interest brought him in contact with one of Austria's famous mineralogist, Gustav Tschermak who was a Professor in Geology at the University of Vienna. Tschermak and Schrauf inspired him profoundly. Becke graduated in 1980 from there and took up various teaching positions during the next two decades beginning with University of Czernowithz in Romania. In 1898, he joined his *alma mater*, University of Vienna, following the retirement of Schrauf, and finally succeeded Tschermak as the Chairman. He rose to the position of the Rector of the Univeristy. He retired at the end of a very successful career in 1927.

Friedrich Becke
(1855–1931)

Becke was a pioneer in optical mineralogy and along with Rosenbusch, Fletcher and others, laid a solid foundation of this branch of mineralogy. Another branch of earth science that was benefited by his seminal contributions was the metamorphic petrology. He served as editor of *Tschermaks Mineralogische und Petrographische Mitteilungen*. He was also made Honorary Fellow of almost all the leading European mineralogical-petrological Societies. For his excellent contribution, Prof. Becke was awarded the Wollaston Medal of the Geological Society of London in 1929.

Prof. Friedrich Johann Karl Becke died on June 18, 1931. The Austrian Mineralogical Society (founded in 1901) awards the Becke Medal, a tribute to the famous mineralogist and their second president, to outstanding scientists for their significant contributions to mineralogy, crystallography, petrology, and other disciplines in geophysics and geochemistry.

The figure reproduced from Michael W. Donaldson: Optical Microscopy at the National High Magnetic Field Laboratory. The Florida State University website with the kind permission.

indices of the grains and the oil or the surrounding medium. If the oil has lower indices than that of the grain, then, the grain would act like a converging lens (Fig. 3.4A), or would act like a diverging lens when the index of oil is more than that of the grain (Fig. 3.4B). The lens effect facilitates formation of Becke Line.

Refractometry

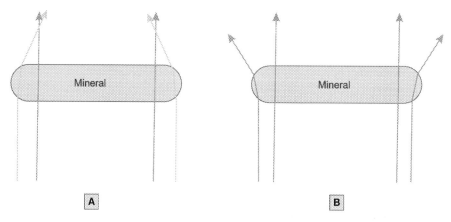

Fig. 3.4 Contribution of 'lens effect' in the formation of becke line. **A.** Converging effect; **B.** Diverging effect.

The *Becke Line method,* or *the central illumination method* is a common method employed in the determination of refractive index of a mineral. The underlying principle is that of the liquid immersion method as indicated in the previous section with a slight modification to help a user to judge about relative order of refractive indices of immersion liquid and the mineral. In order to view the Becke Line, we must bring the grain boundary with the surrounding liquid at the centre of the field of view. It would help if we cut down the unnecessary light by partially closing down the iris diaphragm situated below the stage (Chapter 5, page YY) so that only the desired field of view is brightly seen. In the next step we choose a reasonably low power objective, say, 10X; you will notice the becke line as a sharp light band parallel to the outline of the grain; it may be situated within the grain or outside the grain within the immersion liquid. It may, at times, become necessary to increase the distance between the objective and the object slightly, *i.e.*, defocus the image slightly for the best view of the becke line, because the line is always found in the upper part of the grain as demonstrated in Fig. 3.3. The grain boundary splits into several fine lines, but the Becke line is the sharpest. Now, if the distance between the objective and the object is increased even a bit more, you will notice that the Becke line would appear to move towards the medium of higher refractive index. This technique allows us to judge the value of the refractive index of the mineral with respect to that of the oil. Care must be taken to make sure that the objective movements are very small, otherwise the image will become defocused completely; no useful purpose would then be served in such a case. Fig. 3.5 shows the actual cases of the Becke line movements for the two cases of comparative refractive indices.

Quite commonly minerals do not have nice vertical edges as shown in Fig. 3.3. We all know that a majority of tetragonal and hexagonal minerals have dipyramid or rhombohedral terminations on their prism faces. When the *c*-axis of such a mineral lies parallel to the thin sections, the situation will be more like what is shown in Fig. 3.4. Such cases are also commonly exhibited by those orthorhombic and monoclinic minerals which have habits of first and second order prism terminations, or commonly show elongation along *y*-axis, *e.g.*, epidote. In Fig. 3.6A two grains are shown in contact in a rock in such a manner that the contact is gentle sloping towards right. The mineral above the contact is of higher refractive index than the one on the left (mineral M_1). When you look down the ocular you see two grains

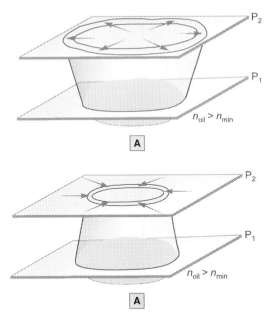

Fig. 3.5 Movement of Becke line as the objective – object distance is increased. P_1 is initial position of the objective when Becke line is observed. P_2 is its second position. **A.** Situation when $n_{oil}>n_{min}$; **B.** Situation when $n_{oil}<n_{min}$.

lying side by side with mineral on the left being of lower refractive index. Suppose you know this mineral, *i.e.*, you can identify this mineral, but mineral on the right is confusing you concerning its identification. It may help if it becomes known whether the mineral on the right is of higher refractive index than the one on the left. Becke line method can

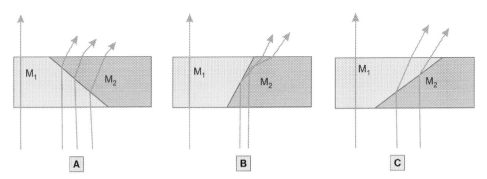

Fig. 3.6 The effect of the inclination of mineral contacts in thin sections on the formation of Becke line. $nM_1<nM_2$. **A.** Becke line may not form; **B.** Becke line is observed; **C.** Becke line may be formed.

Refractometry

easily solve this problem if applied with caution. When the rays of light travel mineral on the left (M_1) first, and, then enter the one on the left (M_2) in a situation depicted in Fig. 3.6A, it will bend towards the normal. You may not see Becke line in such cases; such situations are called *thin sectioning effects* as the ocular view by itself does not give a clue regarding the orientation of the contact. In Fig. 3.6B the contact is steep with a slope towards left so the rays first entering the higher refractive mineral may meet the contact within critical angle; similarly in case of Fig. 3.6C rays will often enter the mineral of higher refractive index and the Becke line law will be fulfilled. In such cases where minerals are anhedral, or have undergone deformation or reactions in post-crystallization periods, all the contacts shown in Fig. 3.6 can occur simultaneously, *i.e.*, the contacts are irregular or toothed so that inclinations are not uniform. You may be surprised to see two or more Becke lines moving in opposite directions. It may be better to shift your observations to another more suitable contact. Alternatively, a slight adjustment of the sub-stage diaphragm may help you to get sharp single Becke Line.

In the above paragraphs we have assumed that the light source is monochromatic. However, it is a common practice to use sunlight for routine microscopic examination, so we must be aware of the dispersion effect on the formation of Becke line. The important point to be kept in mind here is nature of the dispersion curve (Chapter 8, page YY) for the two media in contact. If the curves do not intersect within the visible range then the light will not be effectively dispersed and the Becke line would appear as a band of white light. However, if the dispersion curves of the two media intersect in the visible range, then the color of the Becke line would change depending upon the matching of Becke lines of different wavelengths (Table 3.1).

Table 3.1 Becke line relationships when the microscope stage is lowered (After Nesse, 1985)

Condition	Observation	Interpretation
n_{oil} higher for all wavelengths	White line into oil	$n_{d\,oil} >> n_{d\,mineral}$
$n_{oil} = n_{min}$ for orange red light	Red line into mineral, bluish white line into oil	$n_{d\,oil} > n_{d\,mineral}$
$n_{oil} = n_{min}$ for yellow light	Yellowish orange line into the mineral, pale blue line into oil	$n_{d\,oil} = n_{d\,mineral}$
$n_{oil} = n_{min}$ for blue light	Blue-violet light into oil, yellow-white line into mineral	$n_{d\,oil} < n_{d\,mineral}$
n_{oil} lower for all wavelengths	White line into mineral	$n_{d\,oil} << n_{d\,mineral}$

Oblique Illumination Method

When Becke line is not sharp, or its movement is confusing then *oblique illumination method* is often employed with reasonable success. In this method, sub-stage diaphragm is fully opened and a lower power objective is used. Then nearly half of the field of view is closed by placing a plastic card (credit cards are very useful!) below the condenser, by using one or two fingers, or by inserting an accessory plate half through in the slot. The grain, under this condition, will form a shadow. The shadow will form according to

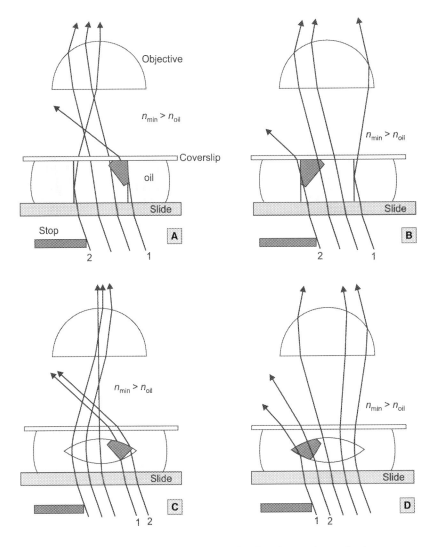

Fig. 3.7 Illustration of the principle of oblique immersion method.

the comparative value of the refractive indices of the mineral grain and the immersion liquid. If the refractive index of the mineral is higher than that of the liquid, then the shadow will form on the opposite side which is being blocked. Shadow will be on the blocked side if the comparative values of the indices are reversed of the first case. Fig. 3.7 gives illustrations of the principle of oblique illumination method. It is apparent from the figure that by blocking a part of the beam we enable the deflection of a large number of rays to such an extent that these do not reach upto the objective lens of the microscope. Here, again, dispersion effect has to be taken into consideration just as we did for obtaining Becke line.

Refractometers

The quantitative determination of refractive index of a mineral is an important step in the systematic description of its physical and optical properties. For an isometric mineral refractive index determination is the easiest as the method neither requires knowledge of crystallographic directions of the mineral nor requires polarized light as mentioned in Chapter 1. We just need a set of refractive index liquids and a *refractometer*. The refractormeter (Fig. 3.8) consists of a light source of monochromatic light ray. The

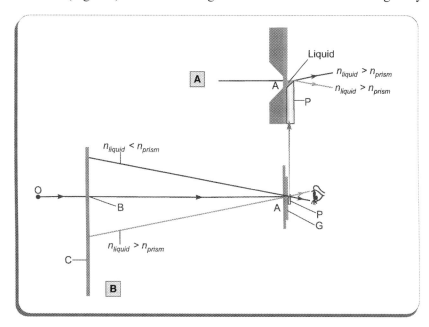

Fig. 3.8 Principle of Leitz-Jelley refractometer.

sodium wavelength ($\approx$ 590 μm; *see* Technical Text Box 3.1) is most preferred. There is a stage on which we put the glass slide. There is a refractometer tube, much similar to

Technical Text Box 3.1

FRAUNHOFER LINES

The Fraunhofer lines represent abosorption spectral lines of solar radiation. William Wollaston noticed this feature way back in 1802 but it was Fraunhofer who provided first systematic description of these lines, hence the name Fraunhofer lines. His careful investigation yielded nearly 600 lines but modern day astrophysicist claim that there are several thousands of these lines in the solar spectrum.

The current thinking is that there are two kinds of these lines. One kind originates in the sun, while the second category of lines are generated when sun rays interact with the oxygen molecules of upper atmosphere. Our interest lies in the lines that originate in the sun; to be precise, we say in the upper layers of the sun. A given element causes a number of lines as it gets abosorbed. Another element produces a second set of absorption lines, and the third element still another one, and so on. Most astronomers follow the original nomenclature introduced by Fraunhofer, although other nomenclatures are also in use. In the Fraunhofer nomenclature strong lines are named by letter A through K and the weaker lines by other letters.

Our interest lies in what is called "sodium doublet" represented by letters D_1 and D_2. The central part of this doublet has the wavelength number 589.29 nm. This wave length is designated as D. This number, incidentaly, approximately lies in the middle of visible spectrum with yellow colour. Because of their well defined wavelengths, Fraunhofer lines are often used to characterize the refractive index and dispersion properties of optical materials.

1. Jenkins, Francis A.; White, Harvey E. (1981), *Fundamentals of Optics* (4th ed.), McGraw-Hill, p. 18, ISBN 0072561912.

that of an ordinary monocular microscope with objective and ocular assembly. In addition there are prisms. The principle on which the refractometer is based, is illustrated in Fig. 3.8. It consists of a small prism of a transparent substance with a refractive index of about 1.70. This value enables measurements of refractive indices in a range of 1.33 to 1.92. The principle of the refractometer is simple. The prism (Fig. 3.8A at P) is located within the instrument in such a manner that it is possible to put a drop of the refractive index oil, or the newly obtained oil mixture within the angle subtended by the prism. The prism is cemented to a glass plate, G, which receives a monochromatic ray of light through an aperture, A. The liquid-prism combination refracts and reflects light so as to form an image on the screen, C, of a slit, B, made in the screen. In Fig. 3.8B, A is the source of monochromatic radiation, most commonly a sodium lamp. Following Snell's Law, an image of the slit formed above it, means that the $n_{prism} > n_{liquid}$; if the image is lower than the slit, then, reverse is the case. By slightly tilting the prism, the two images

Refractometry 43

(one direct and the other refracted-reflected) are brought together, *i.e.*, parallax removed. Calibration of the zero parallax position with the refractive index of the liquid is done by manufacturer and shown as an image scale on screen C from where the observer can directly read it.

The set of refractive index liquid is commercially available, but may also be assembled in laboratory. If you plan to make your own set, then two common liquids are carbon tetrachloride ($n = 1.466$) and methylene iodide ($n = 1.744$). The main requirement for a good immersion liquid is transparency and optical inactivity; preferably it should be colorless or of faint color (Appendix N). Two factors that affect the accuracy of results are wavelength of light radiation and temperature of environs during experiment. Accuracy of 0.001 was achieved by early twentieth century optical mineralogists (*e.g.*, Larsen, Emmons etc.) by varying both the wavelength and the temperature.

The two liquids mentioned above can mix very well in all proportions and the mixture exhibits an intermediate refractive index depending upon the relative proportions of the two liquids. The procedure to measure refractive index is as follows:

1. Take a small transparent grain of the desired mineral on a clean glass plate and observe its relief under air to estimate its refractive index.
2. Then choose the mixture of the immersion liquids (conveniently referred to as 'oil') nearest to the estimated refractive index from the set that you have assembled.
3. Put a drop or two of the oil on to the grain so as to completely surround it.
4. Now, obtain a Becke line and observe its movement to find out which is of higher refractive index between the oil and the grain. As an example, let us assume that the mineral is of higher refractive index than the liquid oil. Then add a tiny amount of the oil with higher index from the next bottle of the set with the help of a clean glass rod.
5. Repeat the Becke line experiment after ascertaining the relief. If the relief has become noticeably fainter than before then add a bit of the same liquid to the mixture till the relief disappears totally.
6. In such a case now, the two substances have the same refractive indices, and we are, now, ready to transfer the new mixture to the *refractometer* for exact determination.

As mentioned earlier in the chapter, early twentieth century mineralogists and petrologists meticulously measured refractive indices of a great majority of minerals. The sources of errors were identified (Technical Text Box 3.2). The method described

Technical Text Box 3.2

SOURCES OF ERRORS IN IMMERSION METHOD

Measurement of refractive index of isotropic minerals is a tedious and time consuming process that often tests the patience of investigators. Anisotropic minerals require orientation that entails accessories that are beyond the scope of this chapter, but certainly demand more time than that required by isotropic minerals. Each step in refractive index determination through immersion method introduces errors, which must be minimized for a reliable result.

In addition to personal source of errors, the first major source of errors comes from the non-homogeneous mixing of the two oils to achieve no-relief situation. Some of these organic liquids are viscous, and it is not easy to mix them well. If there is a delay in transferring the final drop to the refractometer, or sudden exposure to bright light, the evaporation and the light energy may change the index of the mixture. If the dropper used for transferring the fluid to the refractometer is contaminated then a major error would be inadvertently introduced.

We must realize that most oils are calibrated for their index at 20°C. If their storage temperature is not consistent with this value, or exposure to sunlight is long, their would require recaliberation. If the storage was proper but the working temperature is different, then we use the following equation:

$$n_D(T) = n_D + (T_r - T)\left[\frac{dn}{dT}\right]$$

where n_D is the index at temperature T_r and T is the temperature of the oil as given on label of the oil bottle. dn/dT is called temperature correction factor usually given on the label of the bottle.

above is for an isotropic substance; for uniaxial minerals we need to repeat the above method with the plane of polarization oriented with reference to crystallographic direction twice to get the maximum and minimum refractive indices. For biaxial minerals we have to repeat the process three times. For both types of minerals getting a small transparent euhedral crystal always helps. But, more than the crystals you need to know certain concepts peculiar to optical mineralogy, which are detailed in the next chapter.

Refractometry 45

> ## Summary
>
> The grain boundary of a mineral as observed under microscope is called its relief. The sharpness of the relief depends upon the differences between refractive index of the mineral and that of its environment. The environment may consist of epoxy in which the grain is embedded or another mineral grain or grains, or even a given oil. The relief is termed positive when the given mineral has a higher refractive index than its environment, or negative when the index is lower than that of its environment. The adjectives strong, high, prominent are used to indicate a sharp grain boundary; similarly, low, poor, or faint terms used when one can barely see the grain boundary; intermediate term is moderate relief. The nature of refraction at the phase boundary is a combination of two processes; in one the rays undergo total internal reflection and bunch up with other refracted rays, and in the second process they either converge or diverge depending upon relative refractive indices. The physical manifestation is a bright line following the border of the grain. The line is called Becke Line and is always lying in the region of higher refractive index. Comparison of refractive indices can also done using oblique illumination method, which involves partial blocking of the light beam thus, creating shadows on the grains on one side. In both cases, however, the observations are modified by the nature of dispersion by oil and the grain. By carefully repeating the procedure it is possible to decrease the sharpness of the relief of a grain immersed in a given mixture of oils. In this condition refractive indices of both oil mixture and the grain are same. A drop of this oil is placed on the prism of a refractometer that enables us to quantitatively measure the refractive index through elimination of parallax.

Important Terms

Becke line 31

Central illumination method 37

Dispersion effect 39

High or strong relief 33

Immersion method 37

Lens effect 35

Low or faint relief 33

Moderate relief 33

Negative relief 33

Oblique illumination method 40

Parallax 42

Positive relief 33

Refractometer 43

Questions

1. What is relief of a mineral?
2. What are the factors on which relief of a mineral depends?
3. Can a mineral show low relief in one example and high relief in another example?
4. What is lens effect?
5. Explain with the help of diagrams, the formation of Becke line.
6. Briefly explain how the dispersion by oil and mineral control the nature of Becke Line?
7. What is an oblique illumination method for determining refractive index?
8. Explain the formation of shadows in the view of grains during the oblique illumination method for determining refractive index.
9. Describe the principle of Leitz-Jelley refractometer.
10. Discuss the sources of errors in immersion method.

Suggestion for Further Reading

Dowty, E., 1976, Crystal Structure and Crystal Growth: I. The influence of internal structure on morphology, Amer. Mineral, 61, 448-459.

Emmons, R.C. and Gates, R.M., 1948, The use of Becke line colors in refractive index determinations, Amer. Mineral, 33, 612-618.

Hecht, E., 1987, Optics, Addison-Wesley.

Hurlbut, C.S., 1984, The jeweler's refractometer as a mineralogical tool, Amer. Mineral, 69, 391-398.

Laskowski, T.E., Scotford, D.M. and Laskowski, D.E., 1979, Measurement of refractive index in thin sections using dispersion staining and oil immersion techniques, Amer. Mineral, 64, 440-445.

Nesse, W.D., 1986, Introduction to Optical Mineralogy, Oxford University Press, Oxford UK, 325p.

4
Optical Crystallography

Learning Objectives

- Propagation of light in various media
- Concept of indicatrix
- Optic orientation of minerals

Prologue

Sensation of color in substances is one of nature's best gifts to humans. Color in plants, their roots, trunks, leaves, or flowers are pleasure to look at, particularly during the spring. Colors in rocks, gems, or even soils are very delightful to viewers. Many poets and philosophers sought the solace when dawn and dusk mix colors in the sky. Minerals lose much of their colors when thin sections are made. However, using concepts put forward in this chapter we can add colors in such a way to various minerals that is aesthetically pleasing, but, also, allows us to identify minerals easily.

(Courtesy: T. Verma)

A rainbow during a clear afternoon sky in Michigan.

Chapter's Outline

- Isotropic optics
- Uniaxial optics
- Biaxial optics

© The Author(s) 2023
P. K. Verma, *Optical Mineralogy*,
https://doi.org/10.1007/978-3-031-40765-9_4

Introduction

We have already learnt from Chapter One that refractive index of a substance provides us a convenient tool to subdivide materials into different categories that also correspond to different manners of aggregation of atoms within the material. Table 4.1 summarizes what we already know. The last column contains information about two categories of substances that we normally encounter, *i.e.*, amorphous and crystallines. The crystallines are further divided into isometric, uniaxial and biaxial minerals. The nature of variation

Table 4.1 Relationship between refractive index and different crystalline states

Nature of refractive index variation	Name of the category	Crystalline state
No variation in refractive index along x, y, z direction	Isotropic	Amorphous, glass, Isometric crystals
No variation within x-y plane but varies in other direction between two limits, which are characteristic of the mineral	Anisotropic uniaxial minerals	Hexagonal and Tetragonal crystals
No variation normal to two directions that are not necessarily related to x, y, z directions of the crystal. But the index varies in other directions between two characteristic limits; exact characterization requires reference to three mutually perpendicular directions α, β, γ	Anisotropic biaxial minerals	Orthorhombic, Monoclinic and Triclinic crystals

of refractive index within each of these categories is described in column one. You will notice that some of the information given above was not described in Chapter 1. We shall intensively deal with concepts on which this new information is based in this chapter. We begin with isotropic minerals.

Isotropic Optics

In isometric crystals (we shall defer discussion on glass for future) atoms on the unit cell scale are distributed in such a manner that the chemical bonding between atoms is uniform. In other words, electronic environment, energy distribution and other such factors that could influence coherent scattering (*see* Chapter 1) of light photons are evenly distributed throughout x, y, and z directions. Deviations may arise from this ideality. One common cause is the fine-scale internal fracturing and annealing in the mineral. The fracturing causes distortion in unit cell so the energy distribution is no longer uniform, although morphology of the grain may still retain isometric characters. You must have known about liquid crystals that are used in electronic displays. These liquids do not possess long range order but on short range contain strongly aligned asymmetric molecules that are responsible for the display.

The Isometric Indicatrix

Concept of *indicatrix* is due to Fletcher. An indicatrix is a geometric 3-D figure whose origin coincides with that of the crystallographic directions and optic directions form its axes[1]. The lengths of the axes are proportional to the value of the refractive index along that optic direction. In Fig. 4.1 the cartesian axes x, y, and z are shown. Let us assume a ray traversing along x-axis, with its vibrations parallel to z direction. Then the refractive index along z with which the ray propagates along x is n_a. With this refractive index scaled to OZ both the upper and the lower part of the z-axis are marked. Similarly, rays along y and z can be said to mark OX and OY respectively as refractive indices. Here, we are considering an example of isotropic medium where refractive indices are same in all directions; hence, OZ = OX = OY. In other words, $n_a = n_b = n_c = n$ (say). Therefore, for any isotropic mineral the indicatrix will be a sphere with radius equal to n, its refractive index. The size of the indicatrix is determined by the refractive index of the mineral.

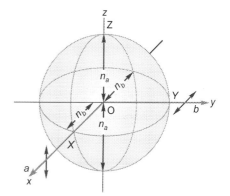

Fig. 4.1 A perspective view of isotropic indicatrix. Note that ray a propagates along x direction with vibration parallel to z direction with a refractive index n_a; similar case exist for y and z directions. Since the three refractive indices are equal, the isotropic indicatrix is a sphere.

Because of the uniform nature of the indicatrix a polarized light vibrating in any direction will not be "pulled away" to a favourite plane. The exact meaning of this sentence will be clear to you when we discuss the optical nature of a uniaxial crystal. Although, the plane of polarization of incident light is unaffected, the ray may suffer otherwise through loss of intensity because of the presence of defects in the crystal structure, presence of foreign ions, or presence of ions of multivalent transition elements. A polychromatic ray may undergo dispersion in addition to absorption. The ray that emerges from the crystal may thus be colored but without any change in the orientation of the plane of vibration.

---

1. An elegant description of background is presented in Technical Text Box 4.1.

Uniaxial Optics

From the point of view optical mineralogy minerals crystallizing in the Tetrahedral and Hexahedral crystal systems are called uniaxial crystals. We have seen in Chapter 1 (Fig. 1.10) how calcite causes double refraction. Calcite crystallizes in Hexagonal system, hence is a uniaxial crystal. It was also pointed out that light traversing along the *c*-axis through this crystal does not undergo double refraction. In the following section this phenomenon is examined in depth.

Uniaxial Wave Surfaces

Huygens' explanation of the phenomenon of two image formation, *i.e.*, double refraction, was that the ordinary travelling with a constant velocity will reach equal distance all around after a given *t* seconds; the locus will be a sphere just like isometric crystals. But for the case of extraordinary ray, it will travel in different directions with different velocities, and will travel to different distances for a give time interval of *t* seconds; the locus of such points is an ellipsoid of revolution. Since, both rays originate from the same point; wave surface of one will enclose the other wave surface. Here, two situations are possible that are depicted in Fig. 4.2. The figure shows two cross-sections; in Fig. 4.2A, a circle encloses an ellipse, whereas in Fig. 4.2B, an ellipse encloses a circle. In terms of ray velocities, the first case denotes ordinary ray velocity is always higher than that of the extraordinary ray. The two exceptions are top and bottom of the sphere where the ellipsoid of revolution is tangent to it. At these two points both rays share the

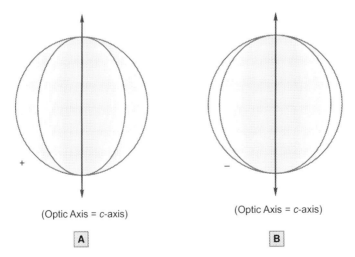

Fig. 4.2 A. A section of wave surfaces formed in uniaxial positive minerals; **B.** A section of wave surfaces formed in uniaxial negative minerals.

same velocity, and define a unique direction, that of the optic axis, coincident with the crystallographic *c*-axis. A particular mineral will show only one of the two behaviours shown in the figure; hence, this behaviour is a characteristic property of the mineral. This allows us to classify uniaxial minerals further as *uniaxial positive* when the ordinary ray velocity is greater than that of the extraordinary ray (Fig. 4.2A), or *uniaxial negative* if reverse is the case. Before we proceed further, it may be instructive to perform an experiment that reveals an important character of anisotropic minerals. Take two calcite crystals that are transparent and approximately of the same size, and view an object as shown in Fig. 4.3A. Calcite rhomb no.1 spilts the light ray into two rays but calcite rhomb no.2 allows the two rays to pass through unchanged. So when you expect four images in accordance with the phenomenon of double refraction but only two rays are emerging at the end of the second crystal. The explanation lies in the anisotropic distribution of electric charge within an anisotropic crystal as different ions are packed in different manners in different directions. Technical Text Box 4.1 gives a tensor treatment of how a momentum induced by the entry of a light photon in an anisotropic medium is resolved in three directions that are mutually perpendicular to each other. Within a given plane the displacement vector will be coincident with the incident force direction along two mutually perpendicular directions; whatever is the direction of incident rays, the location of the two component directions is fixed with reference to the crystal axes. These two directions are shown in Fig. 4.3B as A-B and C-D; these are, in fact, traces of two mutually perpendicular planes. These planes are called *vibration planes of the mineral*. E-E′ is the vibration plane of the polarized light emerging from

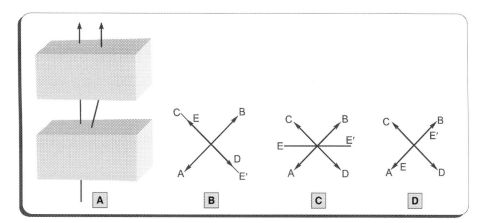

Fig. 4.3 Relationship between vibration planes of a mineral and vibration plane of the incident beam. **A.** Two calcite crystals arranged in parallel orientation. Two rays emerge at the end; **B.** Incident ray vibrates along the plane A-B; **C.** Incident ray vibrates oblique to the planes A-B and C-D; **D.** Incident ray vibrates along the plane C-D.

Technical Text Box 4.1

TENSOR APPROACH TO OPTIC INDICATRIX

Definition of Tensors: Tensors are simply arrays of numbers, or functions that transform according to certain rules under a change of coordinates. A tensor may be defined at a single point or a continuum of points. The latter case is referred to as a tensor field. A scalar quantity, for example, mass of a particle or object, may be thought of as an array of dimension zero or zero order tensor. Density of a fluid as a function of position is an example of a scalar field, or zero order tensor field.

The tensor of order one is known as a vector and one may define it over a field, *i.e.*, a vector field. In ordinary three dimensional space, a vector has three components, x, y, and z; we may write as one dimensional array either as a column or row. An electric field in space is an example of a vector field that requires more than one number to characterize because it has both a magnitude (strength) and it acts along a definite direction. Generally, both the magnitude and the direction of the field vary from point-to-point.

There are several properties that require more than two numbers for proper characterization. These are named tensors of order 2. These are often referred to as matrices, *i.e.*, two dimensional arrays. An example of a second order tensor is the so-called inertia matrix (or tensor) of an object. For three dimensional objects, it is a 3 x 3 = 9 element array that characterizes the behaviour of a rotating body. The response of a gyroscope to a force along a particular direction is generally re-orientation along some other direction different from that of the applied force or torque. Thus, rotation must be characterized by a mathematical entity more complex than either a scalar or a vector; namely, a tensor of order two. The second example is behaviour of thermal conductivity of a substance. This is defined by a relation between heat flow to the temperature gradient. These physical parameters require three principal values for their accurate description.

Before analyzing the behaviour of light, it might be mentioned that elastic properties, gravity etc. are tensors of fourth order requiring 4 x 4 x 4 x 4 = 256 components. Obviously, a very complicated field!

Note in Fig. 1 there are two directions. F denotes the incident force and D is the consequent displacement direction. The component of D along F is $D^*\cos ö$, where $ö$ is the angle between D and F. Since D, or $D^*\cos ö$, is proportional to F, we may write:

Fig. 1

$K = [(D \cos ö)/F]$. ... (Eq. 1)

where K is the constant of proportionality.

Let us analyze F with reference to reference co-ordinates x, y, and z; then the components of F along these are given by:

$F_x = F.l$
$F_y = F.m$
$F_z = F.n$

where l, m, and n are the cosine values of the corresponding angles made by F with the co-ordinate axes respectively.

Contd. ...

Optical Crystallography

Now, we may write the value of D_F in terms of its components;

$$D_F = D_x.l + D_y.m + D_z.n$$

Substituting from above we get the all important equations:

$$K = k_1.l^2 + k_2.m^2 + k_3.n^2 \qquad \qquad ...(Eq. 2)$$

where k_1, k_2 and k_3 are the values of K along the three co-ordinate axes. With the help of this equation we can predict the value of K in any general direction if the values of k_1, k_2 and k_3 are known for F.

Now, let us attempt to relate the terms, F and D with more familiar terms in Optics. In the beginning of this book it has been mentioned that light propagation may be viewed as an electromagnetic wave motion which consists of an electric vector E and a magnetic vector B; also mentioned was the fact that most common practice is to consider more dominant E as a starting point for analysis.

When a ray of light enters a medium, its electric vector E causes a redistribution of electric field creating electric dipoles in the medium. It results in an initiation of electric flux of density D. D will obviously depend upon the value of E via â, permittivity of the medium. In order to explore the nature of âwe may define $â_0$ as the permittivity of vacuum and relate these as follows:

$$K = â/â_0;$$

where K is called dielectric constant.

Then D is related easily to E by the following relation:

$$D = â_0 \, KE \qquad \qquad ...(Eq. 3)$$

It is easy to see similarity in equations 1 and 3; the only difference is that a specific term E has replaced the general term F. Therefore, it follows that the nature of dielectric constant is also that of a second rank tensor and its behaviour is guided by equation 2.

Equation 2 describes the nature of variation of K in three dimension; its ellipsoidal nature is easily understood when we substitute the values of cosines – l, m, and n. cosine l means x/r if r is the scaler displacement vector D and x is its component along x-axis. Similarly, m and n would be y/r and z/r respectively. Therefore, we have for equation 2:

$$K = k_1.x^2/r^2 + k_2.y^2/r^2 + k_3.z^2/r^2; \text{ or}$$
$$r^2.K = k_1.x^2 + k_2.y^2 + k_3.z^2; \text{ substituting } r^2K = 1, \text{ we have}$$
$$k_1.x^2 + k_2.y^2 + k_3.z^2 = 1; \text{ or}$$
$$[(x^2)/(1/\sqrt{k_1})^2] + [(y^2)/(1\sqrt{k_2})^2] + [(z^2)/(1/\sqrt{k_3})^2] \qquad ...(Eq. 4)$$

Equation 4 represents an ellipsoid of revolution with k_1, k_2, and k_3 as its semi-axes. Since, it has been known that for a non-magnetic substance the values of dielectric constant and refractive index are related as:

$$n = \sqrt{K}$$

hence, Equation 4 can be considered as a representation of refractive index variation provided we realize that the refractive index is not the second order tensor, but K is. Since most Optical Mineralogists are familiar with refractive index rather than dielectric constant, it is has become customary to represent the index as a optical response to the induced radiation (incident light ray). If k_1, k_2, and k_3 are equal then it is an *isometric* indicatrix; when $k_1 = k_2$ but not equal to k_3, we call it *uniaxial* indicatrix. Finally, when all three axes are unequal it becomes a *biaxial* indicatrix.

rhomb 1 and is coincident with one of the vibration planes (C-D). The function of the vibration planes is to restrict the vibration of light rays within the mineral. There is no other way light can get transmitted through an anisotropic (uniaxial) crystal except following one or the other vibration plane. If the incident light is plane polarized along A-B (Fig. 4.3B) or along C-D (Fig. 4.3D) then it will pass through the crystal without getting affected because for that crystal these are the preferred planes of vibration. Vibration planes, allowing smooth passage of a ray, do not cause double refraction. However, if the incident plane polarized light enters at an angle (Fig. 4.3C) to these vibration directions, then the double refraction would take place and the incoming light ray will be resolved into two along A-B and C-D forming two images for each rhomb.

Optic Orientation in Uniaxial Minerals

The vibration planes of a uniaxial mineral can be easily specified by means of the limiting values of the refractive indices. The ordinary refractive index, designated as ù, is effective along one of these planes, similarly, extraordinary refractive index, with symbol å, influences the ray velocity for the other plane. For uniaxial positive minerals (Fig. 4.2A) the comparative indices are å > ù, but for uniaxial negative minerals (Fig. 4.2B), these are å < ù. Since, the ray velocity bears inverse relationship with the refractive index, positive minerals will allow light to travel faster along ordinary direction as opposed to the extraordinary plane. It is a common practice in optical mineralogy to designate them as *fast* and *slow directions.* To summarize we may state that every uniaxial mineral will have two directions: one fast and another normal to it as slow. Fig. 4.4 is a *Principal Section*, of an uniaxial mineral, it is named so because it contains the ray and the optic axis, which is the definition of a principal section. The rule is:

1. an extraordinary ray always vibrates in the plane containing the ray and optic axis, and
2. an ordinary ray always vibrates at right angle to plane containing the ray and the optic axis.

The vibrations within the plane of the paper are shown by drawing lines normal to the ray path, and those normal to the plane

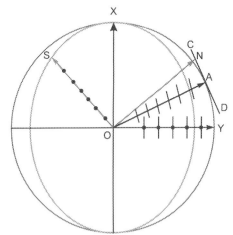

Fig. 4.4 A cross-section of ordinary and extraordinary wave surfaces in a uniaxial negative mineral.

of the paper are marked by bold dots. In Fig. 4.4 OA is the direction of propagation of the extraordinary ray and CD is tangent to the ray surface at A. Extraordinary ray surface being an ellipsoid of revolution, CD will be oblique to OA and not normal. The normal, called wave normal in Chapter 1 is ON. You may note that at Y and X ON and OA will coincide with each other. Another point that should become apparent to you readily is that ON is the reciprocal of the extraordinary refractive index along the ray OA.

Uniaxial Indicatrix

If we draw an extraordinary wave surface (Fig. 4.5) as X-Y with the centre at O, then it is possible to imagine in any random direction a ray as OA with a wave normal as ON. Further, let us draw a line OX′ by extending OX till OX′ becomes equal to 1/OY. Similarly, we may also draw OY′ as equal to 1/OX, which enable us to draw another ellipse X′Y′ concentric with XY. In the next step of geometric construction let us extend OA till A′ where it meets the ellipse X′Y′, and then draw a tangent A′B to the latter ellipse. Now, from O draw a normal to OA′ as OC and

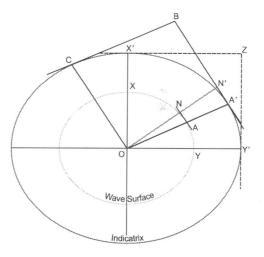

Fig. 4.5 Derivation of a uniaxial indicatrix.

complete the parallelogram OA′BC. By application of elementary theorems of geometry it can be easily shown: OC = 1/ON.

However, it has already been shown that 1/ON is the extraordinary refractive index in the direction of the ray OA. This is an important property of the new geometric construction that is named *Indicatrix*, i.e. each radius of indicatrix represent a unique refractive index. The indicatrix is a three dimensional figure which is associated with refractive indices of a given anisotropic crystal. On the other hand, the wave surface is generated as the light travels through a crystal and, except for one direction, two components of light travel with different velocities. It is hoped that the students will understand this fundamental difference between indicatrix and wave surface.

Fig. 4.6 shows a uniaxial negative indicatrix in a perspective view. It shows that as the light ray OA′ travels, the extraordinary component will vibrate along OC which is the long axis of the elliptical section of the indicatrix normal to ON′. The ordinary

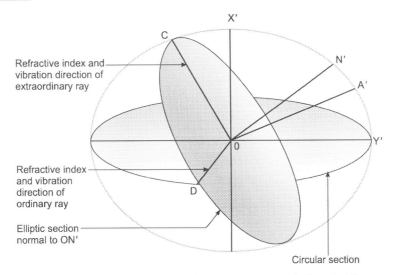

Fig. 4.6 Negative uniaxial indicatrix and its important geometric characteristics.

component will vibrate along OD which is the radius of the circular section of the indicatrix normal to optic axis (OX´).

Biaxial Optics

Biaxial optics is different from the uniaxial optics in several manners. From the treatment given in Technical Text Box 4.1, it may be appreciated that the momentum introduced by an incident light photon in the electric field of a biaxial crystal can be ascertained by resolving it along three mutually perpendicular directions within the unit cell. Two of these directions are along maximum and minimum refractive indices, designated as á and ã (Fig. 4.7) and â is normal to the plane containing á and ã.[6] These directions intersect at the centre of the crystal and lengths OX, OY, and OZ are proportional to the numerical values of á, â, and ã. Since the refractive index distribution is three dimensional within the crystal, it would be instructive if we study its variations in the three standard sections, *i.e.*, á-â, â-ã, and ã-á, individually.

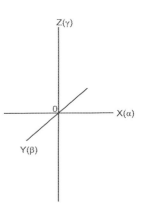

Fig. 4.7 Reference Axes of biaxial wave surfaces.

---

6. In older textbooks, these directions were noted with capital letters X, Y, and Z; the current consensus is in favour of á, â, ã.

Optical Crystallography

Now let us consider a ray of light (Fig. 4.8A) travelling along the plane XOY through O. It will be split in two components; one of them will vibrate along OZ, which is a normal to the plane at O, and the second will vibrate within the plane. The component that vibrates along OZ will do so with refractive index ã, which will have a constant value in the XOY plane; hence its wave surface is a circle within this plane, and the circumference is a locus of point where this component of the ray has reached in a given time (compare with the uniaxial wave surface). Since the second component is vibrating within the plane, the location of points where the ray has reached in the given time will be governed by the direction with reference to the centre O. If the ray direction is OX, it will vibrate along OY and is associated with refractive index â; similarly, if the ray is propagating along OY, the vibration is along OX and the refractive index is á. Hence, the locus will be an ellipse that will enclose completely the circular locus of ã because ã being the highest refractive index will propagate light slowest.

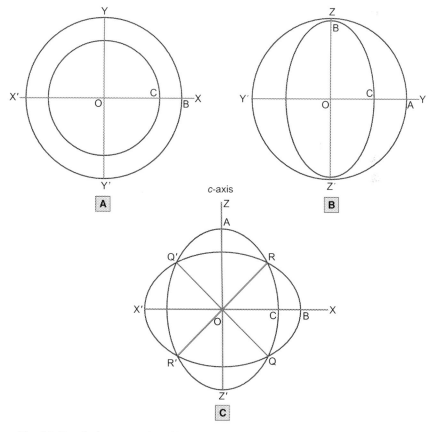

Fig. 4.8 Standard cross-section of biaxial wave surfaces. **A.** Along á – â plane; **B.** Along â – ã plane; **C.** Along á – ã plane.

Similar arguments will hold good for the construction of wave surfaces in YOZ plane (Fig. 4.8B) except that now circular locus representing velocities associated with refractive index á will completely enclose the elliptical wave surface.

However, the geometrical relationships of the wave surfaces are distinctly different in the XOZ plane (Fig. 4.8C). The refractive index normal to plane is â which is more than á, but less than ã. So the circular locus will intersect with the elliptical one with important consequences on the velocities of the two rays. The two surfaces intersect at four points implying thereby that at these four points the two velocities are coincident; compare this with the condition of tangency in uniaxial wave surfaces. These four points may be joined along two directions ROR´ and QOQ´. Interpreting in physical terms, it means that a ray passing through O along any of these two directions, called *biaxial optic axes* or, *optic biradials*, will have both components travelling with the same velocity, or in other words, do not suffer double refraction.

It must be emphasized here that when a ray of light is transmitted through a biaxial crystal our assumption of its getting split into two rays is different from the concept of the uniaxial minerals. In the present case, both rays are extraordinary and the circular sections are apparent. Likewise, the two directions (optic biradials) are not optic axes in the sense of uniaxial crystals, although the apparent behaviour of light along these directions is similar.

The Biaxial Indicatrix

Fig. 4.9 is an illustration of a biaxial indicatrix that is a triaxial ellipsoid meaning thereby that any section through its centre will reveal an ellipse, except two sections shown by dashed outlines that are circular. The semi axes, *i.e.*, OX = á, OY = â, and OZ = ã by construction. A random section of this figure through its centre O will have two semi axes; one of them will be shorter than the other, and are proportional to refractive indices of two wave fronts which propagate normal to the given section. However, in the case of the two circular sections which have same radius equal to OY, the ray propagating normal to either of them will have the two components vibrating with the same frequency that is proportional to OY. These two directions are ROR´ and QOQ´ which we have defined in the last section as the optic axes or biradials. The angle

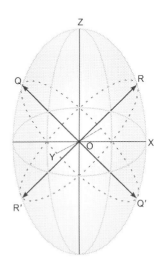

Fig. 4.9 The Biaxial Indicatrix.

Optical Crystallography

ROQ is called the *Optic axial angle*. Its complimentary angle, ROQ' is also called *optic axial angle*. To avoid confusion, the angle ROQ with OZ as bisectrix is abbreviated as $2V_{a'}$ and the angle ROQ' is abbreviated as $2V_{a}$. The plane that includes OZ, OX, and the two optic axes is called *Optic Axial Plane* (OAP). The numerical value of $2V_{a'}$ and the relationship between $2V_{a'}$ crystallographic axes, a, b, and c and á, â, ã, the optic directions, i.e., the location of OAP within the unit cell, is highly sensitive to the configuration of the electric field within the unit cell. Hence, often a minor substitution of an element by another brings about changes in the value of $2V_{a'}$ and/or the orientation of OAP. It was no surprise that a great part of the 20th century was devoted to correlating elemental substitution with OAP location and $2V_{a'}$ values. Best known examples are olivine and plagioclase.

The value of $2V_{a'}$ is used to classify biaxial minerals as positive when the value of $2V_{a'} < 90$; if $2V_{a'} > 90$ then the mineral is termed as the biaxial negative. The two figures given in Fig. 4.10 graphically illustrate this + and − characters of biaxial minerals.

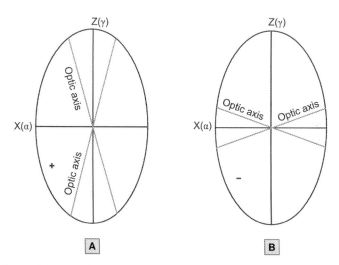

Fig. 4.10 A. Principal Section of a biaxial positive mineral; **B.** Principal Section of a biaxial negative mineral.

Optic Orientation of Biaxial Minerals

The *optic orientation* of a biaxial mineral is partly a property of the crystal system and partly depends upon the electric field of the atoms of the unit cell. The system dependent characters are as follows:

(a) **Orthorhombic System:** All orthorhombic minerals are biaxial and the crystallographic axes (*a*, *b*, and *c*) coincide with optic directions (á, â, and ã). However, it must be borne in mind that which of the crystallographic axis coincides with which of the optic direction is to be determined by an actual experiment. In other words, for a given mineral whether *a*-axis will coincide with á, â, or ã is to be determined by us by experiments that we shall describe in a later chapter. In part II of this book a statement and a diagrammatic representation of this relationship is given for each biaxial mineral. For common minerals of your day-to-day experience you may well be advised to learn by heart this relationship.

(b) **Monoclinic System:** The *b*-axis of monoclinic minerals is a unique axis in the sense that it is the only direction which is an axis of symmetry (two fold) in a monoclinic crystal. In few cases this symmetry is absent but a plane normal to *b*-axis is present as the only mirror plane. This character is also reflected in the optic orientation as one of the three optic directions, á, â, or ã, coincides with *b*-axis of monoclinic crystals. As outlined above, the specific coincidence is to be determined by actual experiment.

(c) **Triclinic System:** Crystal axes of this system have no predictable relationship with the optic directions, therefore, á, â, or ã make acute angles with *x*, *y*, and *z* crystal directions. Hence, optic orientation of a triclinic crystal consists of a statement of these angular relationships.

Summary

Isotropic minerals crystallize in isometric system and are characterized by uniform chemical bonding along all three directions causing identical behaviour of the light ray along these directions. Hence, an isometric wave surface is a sphere. On the other hand, anisotropic minerals are classified as uniaxial and biaxial minerals. Uniaxial minerals exhibit two wave surfaces: a sphere representing ordinary ray wave surface, and an ellipsoid representing extraordinary wave surface. Both are tangential at two points that represent the optic axis along which no double refraction take place. The biaxial minerals also generate two wave surfaces, but both surfaces are extraordinary surfaces and intersect at four points. These four points lie along two directions which are called optic axes and define an angle called optic axial angle (2V). Various optic characteristics of anisotropic minerals are best represented by a geometric figure of ellipsoid called indicatrix. In uniaxial minerals indicatrix is constructed by taking semi-axes as equivalent to å and ù; whereas for biaxial minerals the semi-axes represent values of á, â, and ã.

Important Terms

Biaxial optic axes 58

Biaxial indicatrix 58

Biaxial wave surface 57

Biaxial positive 59

Biaxial negative 59

Indicatrix 49

Optic axis 60

Optic axial angle 59

Optic biradials 58

Optic axial plane (OAP) 59

Optic orientation 59

Optic orientation in biaxial minerals 59

Optic orientation in uniaxial minerals 54

Principal section 54

Uniaxial negative 51

Uniaxial positive 51

Questions

1. Describe the path of light as it originates within a fluorite (isometric system) crystal and propagates outwards.

2. Is propagation of light in fluorite different from that in quartz (hexagonal)? Explain.

3. Is propagation of light in quartz different from that in olivine (orthorhombic system)? Explain.

4. What is Fletcher's Indicatrix?

5. What are essential differences between indicatrix and wave surface?

6. Describe uniaxial indicatrix.

7. Explain the relationship between optic axes and biaxial wave surfaces.

8. What is optic axial angle? Explain its importance.

9. What is a principal section of a uniaxial mineral?

10. What is a principal section of a biaxial mineral?

11. Give optic orientation of biaxial minerals.

Suggestion for Further Reading

Harthsorne N.H. and A. Stuart, 1969, *Practical Optical Crystallography*, American Elsevier Publ. Co. Inc.

Nesse, W.D., 1986, *Introduction to Optical Mineralogy*, Oxford University Press, Oxford.

Whalstrom, E.E., 1964, *Optical Crystallography,* 5th ed., Wiley, New York.

■■■

5
The Polarizing Microscope

Learning Objectives

- Object image relationship in a compound microscope
- Peculiarities of a polarizing microscope

Prologue

He continuously pestered his Nanny to let him cook. He was confident that he knew how it was done. Easy! Not a baby anymore he was nearing three; he could talk and sing, so what was there in cooking! The poor lady was almost in tears out of frustration. Suddenly she heard a commotion outside in the street. A bioscope man with a Santa Clause Bell in one hand was inviting young and old to see "Taj Mahal, Gateway of India, A fat lady, A Prince….." The Nanny took the little boy out, and asked the bioscope man to start the show. The man obliged, took a small folding chair from his bag and adjusted it in front of one of the gazing tubes. Seating the child on the chair he helped tiny hands to clutch the tube and cajoled him to peep into the tube. Ah! The child's face brightened up with immense pleasure as a whole new colorful world opened before him to explore!

— Childhood Story

(Courtesy: A. Verma)

A brightly colored butterfly and a flower on a bright spring morning.

Chapter's Outline

- The compound microscope
- The polarising microscope
- Use and care of polarizing microscopes

© The Author(s) 2023
P. K. Verma, *Optical Mineralogy*,
https://doi.org/10.1007/978-3-031-40765-9_5

Introduction

The central equipment in optical mineralogy is the polarizing microscope, also named as petrologic or petrographic microscope. It is essentially a compound microscope with provision for optical studies of minerals. By now, you have learnt basic principles of optical crystallography; hence, you are now capable of appreciating the special features that are added to a compound microscope. You shall realize, as we proceed in this chapter, the need of these features as well as their characteristic design. The present chapter describes what is known as the basic unit; however, many optical properties described in this book would require the use of supplementary equipments or attachments, which are known as accessories, hence should be ordered while procuring the basic unit[1]. Some accessories are needed only for some specialized examination, and its requirement may or may not arise for a particular laboratory.

The Compound Microscope

The single lens microscope has the disadvantage of being of awfully small size, and therefore, very inconvenient in use. The compound microscope became popular about 200 years ago once various aberrations, introduced by multiple lenses, were reduced. A polarizing microscope is a compound microscope with provisions for special parts that allow us to generate polarized light and analyze optical properties of a crystal. Hence, the first step in understanding the working of a polarizing microscope is to have good grasp on various components of a compound microscope.

Design of a Compound Microscope

Fig. 5.1 shows a sketch of essential features of a compound microscope where the source of light is located near the bottom of the instrument. The source of light can be a plane reflecting mirror that receives sunrays. Very commonly, a halogen lamp is used as the source, which is regulated by a transformer for controlling the intensity of light emission. There is slot for putting filters in the path of light; normally a ground glass to reduce the glare, and a day-light filter are placed in the slot for best results. In high end microscopes a centering device is also attached so that illumination is well-centred and uniformly spread throughout the field of view. There is an *iris diaphragm* (Id) provided close to the source of light for controlling the size of the beam. It normally consist of an arrangement of plastic leaves so that nearly a pinhole is formed when the diaphragm is fully closed. A lens assembly, called the *condenser*, is situated above the diaphragm but below the stage at (C). The condenser assembly may consist of upto three components. The simplest assembly is with only one component consisting of a combination of two

---

1. Many vendors quote various parts of the basic units separately since they offer choices of these parts with different cost implications. Beware that you are ordering the entire basic unit plus some other accessories without which the optic examination would remain incomplete.

The Polarizing Microscope

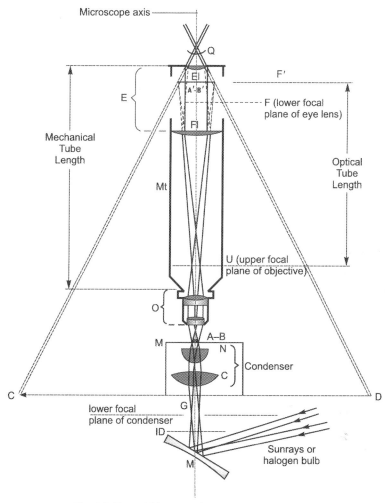

Fig. 5.1 Essential features of a compound microscope.

lenses of an effective aperture of about 0.22. Many microscopes have provision of another assembly of lenses on a movable mount which can be introduced at will when high power objectives are used. Still, some high-end microscopes have provision for a third condenser to be used with oil immersion objectives. Then moving up you may notice a stage with a central hole; this is called the *microscope stage* (M) which has hole in the centre. A thin section (prepared according to the procedure outlined in Chapter 2 (page YY) is placed on this stage so that this hole is covered by the thin section. Directly in line and above the hole is the *objective* (O). The objective is also an assembly of lenses, which, as will be shown later, govern the resolution of the microscope. The objective is held in a tube called the *microscope tube* (Mt), whose other end carries the *eye piece* or the *Huygens Ocular* (E).

Passage of a Light Beam across a Compound Microscope

The objective forms a real image of an object, say, a mineral grain A-B (Fig. 5.1), somewhat enlarged and inverted, at A´-B´ nearly at the top of the tube; its exact location is a function of the focal length of the objective. However, the lower lens of the eye piece, termed the *field lens* (Fl) intercepts the rays and forms an image at A´´-B´´ nearly coinciding with the lower focal plane of the *eye lens* (El) of the eye piece. If the eye is placed at Q, where the rays cross, then a highly magnified virtual image is seen as C-D below the condenser as enlarged right side up.

It is apparent from Fig. 5.1 that the combined magnification of the microscope would largely be governed by the ocular, which, in fact, is the case. However, it is important that we discuss the issue of magnification at this point. There is a general perception that the magnification of a compound (also polarizing) microscope is obtained by multiplying the values engraved on the objective and the ocular. For example, an objective may have a value engraved 10X, 50X, or 40X, while the ocular values may be 5X or 10 X. Therefore, using an objective of 10X and ocular of 5X ought to give a magnification of 50 for image of A-B in Fig. 5.1. This is approximately true, although taking a photo (digital or chemical), and, then, writing a product as magnification is erroneous as C-D, being a virtual image, does not produce a film image. Equations and figures given in Technical Text Box 5.1 clearly outline the calculations involved in obtaining the magnification.

If g is the distance between upper focal plane of the objective and the image formed by the objective, *i.e.*, A´-B´, the magnifying power of the objective is equal to g/f_o, where f_o is combined focal length of the objective assembly. From the figure it is obvious that for practical consideration g should be less than the length of the microscope tube. We have some freedom in choosing the length of the tube but not much as too short or too long tubes would be highly inconvenient to the user. Hence, in a way the magnifying power of the objective can only vary between limits imposed by the relation of its g and the tube-length.

From the Technical Text Box 5.1 we learn that the magnifying power of the eyepiece or ocular is equal to D_v/f_e where f_e is the focal length of the ocular combination, and D_v is least distance of distinct vision. The ocular, known as the Huygens' ocular, is characterized by the absence of focal plane on the field (lower) lens side (Fig. 5.1). The eye lens of the ocular picks A´-B´, the image that would have formed in the absence of eye lens, and forms image A´´-B´´ on the same side. Now, we may write:

$$\frac{\text{Size of A}''\text{-B}''}{\text{Focal lengh of eye lens}} = \frac{\text{Size of A}''\text{-B}''}{f_e}$$

This leads to

$$M = \frac{g}{f_0} \times \frac{D_v}{f_e}$$

Technical Text Box 5.1

PROPERTIES OF LENSES

Let us imagine a biconvex lens with similar curvature on both sides, and let its optical centre be at O (Fig. 1A); this means that any ray passing through O will not be bent in its path. Furthermore, let AOB be its principal axis. Following Snell's Law, we may say that an object at A will form the image at B on the other side of the lens. The distance AO is, say, of length u, and OB is v. For the sake of simplicity, let us assume

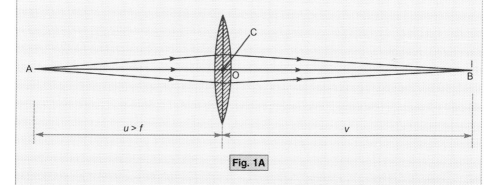

Fig. 1A

that the lens is very thin as compared to u and v, and the rays emanating from the objects are of one wavelength as well as make a very small angle with the principal axis. If u is decreased by a small amount, then, we find that v increases; if u is progressively decreased, then, for a particular value of u, v will become infinity, i.e., the rays will emerge as parallel rays on the other side of the lens (Fig. 1B). This value of u is denoted by f, the focal length of the lens. When $u < f$ then no real image

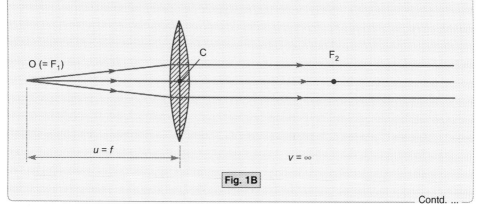

Fig. 1B

Contd. ...

is formed though one would see a virtual image on the same side of the lens as shown in Fig. 1C. The well known fundamental lens equation is as follows:

$$\frac{1}{u}+\frac{1}{v}=\frac{1}{f} \qquad ...(1)$$

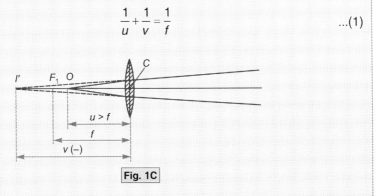

Fig. 1C

In Figs. 2A, B, and C the comparable situations to those in Fig. 1 are sketched but with a finite size of the object. Taking help from the theorems of geometry, it is easily shown that the following relation holds true:

$$\frac{O}{I}=\frac{v}{u}=M=\frac{g}{f} \qquad ...(2)$$

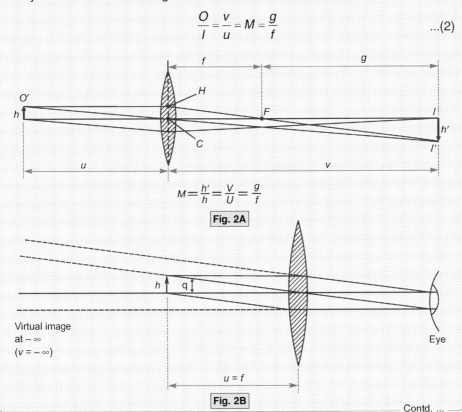

$$M=\frac{h'}{h}=\frac{V}{U}=\frac{g}{f}$$

Fig. 2A

Virtual image at $-\infty$
($v = -\infty$)

Eye

$u = f$

Fig. 2B

Contd. ...

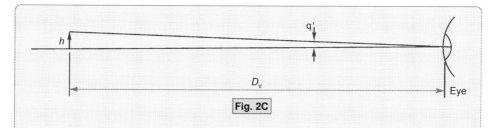

Fig. 2C

Where O is the size of the object, I is the size of the real image formed, M is the magnification of the lens, and g is distance from the second focus to the real image formed.

Now, in using the microscope we always bring our eyes close to the ocular. Image formation at the retina of the eye when the object is at the focus is shown in Fig. 3. In this case the eye would see an infinitely distant virtual image of the object. In this case, the case of a hand lens that you carry in the field, the magnification is given by ratio of the size of the virtual image and that of the object viewed without the lens at the least distance of vision, D_v, normally taken as 25cms. Therefore, the M is given by:

$$M = \frac{Dv}{f} \qquad \ldots(3)$$

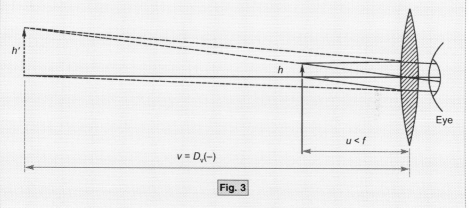

Fig. 3

However, if the object is brought to a position of less than focal length, then the equation (3) is modified to:

$$M = \frac{Dv}{f} + 1 \qquad \ldots(4)$$

indicating an increase in magnification in this case. But there is a practical limit to this case. Closer the object, the divergence between the rays becomes more, and, so a situation would arise when the eye cannot focus on the image.

When you read the design of the compound microscope in the text of this chapter, you would notice these equations are utilized in the construction of the microscope. A real image is formed by the objective at a place that is closer to the ocular than its own focal length. This causes a virtual but enlarged image on the same side of the lens.

which is similar to multiplication of the magnifying power of objective and that of the ocular. As remarked earlier, most manufacturers supply these in round figures of magnification with number followed by 'X' symbol, *e.g.*, 10X.

The Polarising Microscope

The first marked difference between a common compound microscope and polarizing microscope is incorporation of a device to generate plane polarized light (Technical Text Box 5.2) in such a manner that it is possible to control the orientation of this plane. Another difference is in the construction of the microscope stage which is circular, graduated and rotatable with a vernier attached to give readings about the angle of rotation. Then, the stage contains many holes to hold accessories to the microscope. In addition to all these, there is another polarizing device and another set of lenses for various special optical tests that are described in the following pages. In addition, a slot is provided above the objective for inserting accessories to the microscope whenever

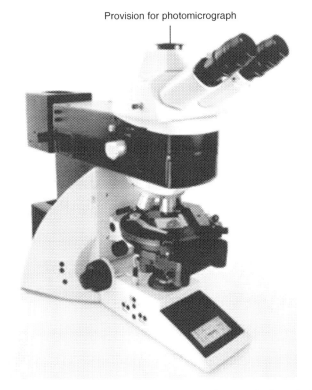

(Courtesy: Leica Corporation)

Fig. 5.2 A photograph of a modern research polarizing microscope.

needed. Various manufactures have different designs and different models to suite needs of users of various levels of efficiency. In Fig. 5.2 a picture of a modern high level research microscope is shown and a line diagram in Fig. 5.3 lists essential parts of another model but you may appreciate that it primarily describes features for most common model. Please note this particular microscope has arrangement for study of polished sections as well, but that is outside the purview of the present text. It is possible to acquire a microscope that does not have this dual arrangement so as to save on the cost.

The use of some of the components shown in Fig. 5.3 requires a background that shall be covered in the following chapters, yet a beginning may be made with the concepts already introduced. The first is the source of light that is shown to be a bulb connected to the mains via a transformer. In South Asian countries where there is abundant sunlight a student model of polarizing microscope uses direct sunlight from a window. Irrespective of the source of light, whether natural or man-made, we must remember that mineralogic study requires sunlight or a beam of light that imitates sunlight. Properties of minerals listed in any standard textbook are based on this assumption. Hence, artificial light would need to pass through what is known as day-light filter. It is a circular disc of blue shade glass of about 5mm thickness usually placed above diaphragm. The iris diaphragm should be adjusted to cut down superfluous light.

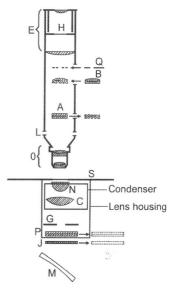

Fig. 5.3 A line diagram illustrating important features of a polarizing microscope.

The light rays, then, enter the condenser assembly that also contains a polarizing device that we shall call the *Lower Polar, Lower Nicol* or *Polarizer*. The entire assembly can be adjusted by screws so that the light rays can be focused on the object placed on the microscope stage. The stage can be rotated freely and engravings allow monitoring of the angle of rotation with the help of a vernier. In many models a screw lock and a clamp are provided to guide the rotation of the stage.

Once the light is through the transparent object place on the microscope stage, it enters the objective. The objective is achromatic as discussed in Technical Text Box 5.3. It consists of several lenses in a stainless steel housing (Fig. 5.4) that has provision for attachment to the microscope tube either directly or through a nosepiece that has several holes in which the objectives may be screwed in. The objective is required to perform two tasks, *viz.*, magnification and resolution of the object. From Technical Text Box 5.1, we know that the magnification is controlled by its focal length.

Technical Text Box 5.2

POLARIZING DEVICES

A general name – polarizer, is given to a transparent medium through which a beam of light is refracted, and in doing so, acquires a certain degree of polarization. The incident beam may or may not be polarized. Well-known mineralogist, A.F. Halimond preferred to call these devices as polars; a plethora of names exists for polarizers because of wide range of material used and techniques employed. In general, we may categorize the polarizers in two categories, viz., absorptive polarizers and beam splitting polarizers.

Edwin H. Land is credited with developing a transparent sheet of nitrocellulose polymer, coated with tiny crystals of iodoquinine sulphate, also known as herapathite. The crystals follow an alignment; normally the long axis of the crystals point in one direction. The light rays follow this direction, causing a flow of electrons along that way colliding with other particles and re-emitting the light along the path of incident ray; this causes cancellation of the incident waves. The light polarized parallel to the crystal alignment is absorbed whereas that part, which vibrates normal to the alignment, is transmitted. This discovery made in 1929 was patented as *Polaroid*. There has been tremendous advancement in technology in Polaroid films increasing their durability and practical applications. Most present day microscope manufacturers use such films, which are glued a glass mount. For specialized microscopes like infra-red microscopes, and also in optical fibre communications, nanoparticles of silver tubes embedded in a glass plate as absorbing polars.

In the early nineteenth century, William Nicol, a Scottish scientist, devised a prism almost 100 years earlier (1828) to the patent date of polaroid that became known as *nicol prism* for obtaining a polarized light. It was the first polarizing device using the beam splitting principle. A transparent crystal of calcite, var. Iceland spar was used. The dichroism of calcite, tourmaline and a few other minerals had been known for quite sometime. The phenomenon was explained by Huygens as splitting of light ray in two components of ordinary and extraordinary (e-ray) polarized as plane polarized rays. Hence, Nicol used a calcite rhomb and filed its edges in such a manner that the angles as shown were reduced to 68° (*see* Fig. 1). Then, he split the

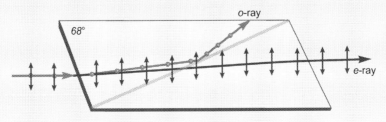

Fig. 1 Construction of nicol prism.

Contd. ...

rhomb in two parts along one of the diagonals. The two halves of the rhomb were then rejoined with the help of an organic adhesive, commercially available under the trade name of Canada balsam. It is thick colourless liquid but becomes solid once heated slightly and then cooled. The solidified cement is isotropic and has refractive index of ~1.55. In order for the split rays to sufficiently diverge, the device is made long and slender. When both split rays reach the cement (*i.e.*, Canada balsam) the angle of incidence is such that (by the design of the device) o-ray meets the cement layer at angles larger than its critical angle value causing it to undergo total reflection. On the other hand, e-ray would suffer refraction and continue to travel in the same general direction of incident ray. The o-ray would get absorbed by the walls of the prism and its housing. The nicol prisms found applications in the polarizing microscope and other instruments. A number of variations of nicol prisms were suggested for specific purposes. Many of these, for example, Glan-Thompson prism, Glan Focult prism, Glan-Taylor prism, Wollaston prism etc. are in use. In comparison to Polaroid filters the beam splitters have drawback in having a very small field of view, and rather inconvenient length of the device. It is for this reason that once, Polaroid filters became available, beam-splitting prisms were no longer commercially popular.

Some manufacturers also engrave the focal length on the lens housing but most supply the magnification number. It is suggested that every time you begin microscopic examination, you should initially use the minimum magnification objective. The resolving power of an objective is related to the separation

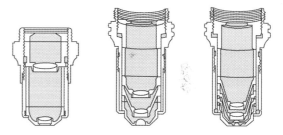

Fig. 5.4 A cross-section of an objective.

of two closely placed points on the object. In order to appreciate this let us look at Fig. 5.5. The *limit of resolution* is dependent on wavelength, (also on illumination to some extent), and also on the angle of the light cone subtended by the aperture of the objective with object as its apex. Let X be the object as shown in Fig. 5.5. The objective is situated at O at a distance L. Light rays emanating from X will form a cone with $\angle OXP = \mu$, *i.e.*, inverse of $5n\sin\mu$, which is called *numerical aperture* (NA). The NA value is also engraved on the objective housing. This is to remind you that lower the limit of resolution more information about the object become available. So a large cone angle is desired for which a small focal length is suitable. Normally an objective of

Technical Text Box 5.3

ABERRATIONS PRODUCED BY LENSES

The image formation processes as described in Technical Text Box 5.1 represent highly simplified processes that are not achieved in a common instrument. In practice, image formed by lenses suffer from aberrations. There are various kinds of aberrations that affect the efficiency of image formation.

Spherical Aberrations: It arises because of a large number of rays being non-paraxial to the principal axis of the microscope. Because of this, each ray has its own focus; as the divergence of the rays increases, the foci are brought closer to the lens. Hence, a blurring effect is seen in image.

Chromatic Aberrations: Since each ray has its own refractive index, each ray will form its own image. For example, as compared to the red ray, blue ray will form image nearer to the lens. It may be emphasized that chromatic aberration is suffered by paraxial rays also.

Astigmatism: Since large part of objects lie outside the principal axis, and the effective curvature of the lens at different points is different, the image formed is drawn out in two perpendicular directions. This aberration is named as the astigmatism.

Distortion: The image of objects, particularly large objects, suffers from distortion as the magnification of outer parts is different than that of the inner parts of the image.

Curvature of Field: This aberration causes the image lies on a curved surface rather than on a plane.

Removal of Aberrations: By choosing a combination of lenses of different shapes, e.g., a double convex lens with a double or plano-concave lens spherical aberrations can easily be reduced or eliminated. Likewise, if these lenses are made up of two different materials such that one of them is of high dispersion, and the other of low dispersion, chromatic aberrations can be reasonably reduced. It may be pointed out that the corrections for the lens aberrations for a microscope objective are carried out on the basis that the size of the object is much smaller than the aperture of the objective. Such objectives are of two categories. One called achromatic objectives use a pair of doublets as shown in the text of this chapter. The other is apochromatic objectives where a combination of fluorite and glass lenses are used. In the high power objective, where spherical aberration is prominent, theory of conjugate object and image points is utilized. According this principle, in some lenses it is always possible to find two points for which there is no spherical aberration. These objectives consist of hemispherical lens. Very high power objectives that are marked 100X or so, the space between these lenses and the object is filled with given oil drops with a refractive index same as that of the lens material. This enables a wide cone beam, as opposed to paraxial rays, to enter the lens without suffering aberration.

about 100X (NA = 1.30) would give a very high resolution, particularly if drops of oil cover the space between the object and the lens. The oil ($n = 1.5$) also reduces the spherical aberration.

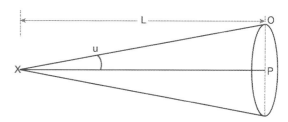

Limits of resolution = $\lambda/(5n\sin\mu)$ = 1/resolving power n = refractive index of the medium separating object (X) and objective (O)

Fig. 5.5 Resolution of the objective.

The light rays after traversing the objective would pass through a number of devices at the discretion of users (we shall learn about them later) and finally enter the eye piece or ocular. The most common type is known as Huygens ocular after the great Dutch physicist (*see* General Text Box 1.1). It consists of two lens systems known as field lens (Fig. 5.4) and eye lens (Fig. 5.4); as the names suggest, eye lens is closer to eye, *i.e.*, it is top lens, and the other one is at the bottom of the ocular (Fig. 5.6). The ocular may be taken out and replaced. At times, it may be desirable to observe the image of the object without ocular. In some microscopes, the camera is fitted directly at the top of the microscope tube after removing the ocular, though such models are now becoming obsolete. Due to a reasonable distance between the objective and the ocular, the latter can receive only such rays that form small angles with the microscope axis. The focal length of the field lens is larger than that of the eye lens by an amount somewhere between 1+1/5 to 3 times. Distance between eye and field lenses is approximately half of the sum of the focal lengths of these two individual lenses.

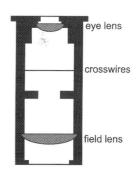

Fig. 5.6 A cross-section of Huygens' ocular.

The idea behind the setting of these two lenses in the Huygens ocular is to ensure that the combination does not have external focal plane on the field lens side. Hence, in order to view crosswires (Fig. 5.6) etc. these should lie on the lower focal plane of the eye piece. In many microscopes, eye lens can also be focused. This is particularly true for binocular microscopes such that shown in Fig. 5.7. This particular model has a provision of a prism system to deflect part of the light to a vertical tube where a camera (digital or chemical) is mounted almost permanently for the convenience of user. In image analyses systems a cable from this tube is connected to a personal computer or a Mackintosh computer.

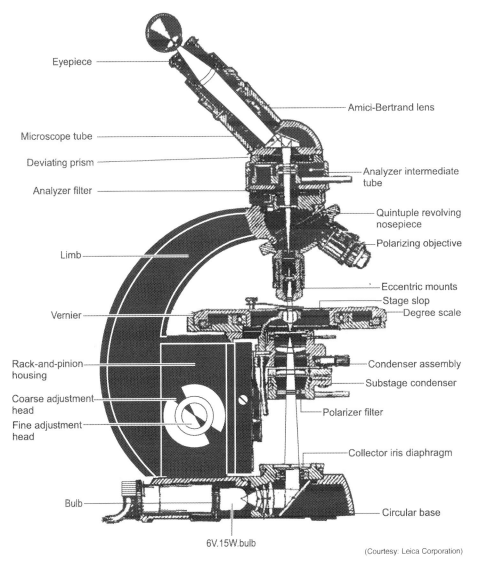

Fig. 5.7 Binocular Polarizing Microscope.

Use and Care of Polarizing Microscopes

The polarizing microscope is a basic tool for geologists as it helps to confirm mineral and rock identification done provisionally in the field. It allows one to estimate chemical composition of minerals that show variation in composition; through textural study microscope permits us to predict the petrogenesis, and furthermore, it guides us in selecting right samples for chemical analyses and spectroscopic analyses. Time spent on detailed microscopic examinations saves time and money on modern sophisticated equipment through right sample selection. Hence, it is not surprising to find many well known scientists over-zealously and over-protectively guarding their microscopes! No wonder that the polarizing microscope term is synonymous with *petrological microscope*.

General Care

The two major enemies of a petrological microscope are humidity and dust. Therefore, locate your microscope in relatively dust free and dry corner of a room. This is easier said than done as often laboratory buildings are designed for general use. Strong hygroscopic material may be kept to absorb moisture if it can not be otherwise avoided. No good worker allows a help to clean his equipment even if he is loaded with project money! He does it himself as he takes charge of work every morning; a simple wipe by tissue paper for 2-3 minutes is all that is required. Multi-user microscopes, like that in an undergraduate laboratory, should be looked after by technicians on yearly contract basis and initially the young learners should not be allowed to clean the equipment.

A step that may occasionally be needed is oiling of certain components. One is fine and coarse focusing device; condenser assembly focus and the rails on which either microscope stage or the microscope tube moves up and down for focusing the image of the object are other parts that may need a small drop of machine oil or oil recommended and supplied by the makers of the equipment. This is required when you are using the equipment after a long gap or putting the microscope back to its cabinet in anticipation of a long period of other schedule. We have already mentioned that one of the peculiarities of the petrological microscope is revolving nature of the microscopic object stage. It revolves through a ball bearing mechanism where the balls are embedded in a silicone jelly. The revolution is finger touch; a resistance encountered is normally due to the locking screw being in place. Hence, no extra effort should be made till the stage lock is examined and released. The resistance also arises because of progressive hardening of the grease of the ball bearings over which the stage revolves. In such an eventuality, only a trained person should unscrew the stage, take out all the balls, clean the channel, and refill the grease replacing the balls and the stage back in place. A regular yearly check-up and cleaning would not allow such things to happen as hinted above.

While using the microscope, one must begin by using the lowest power objective with *upper polarizer* (analyzer) out of the path of light. Bring the image to focus with the help of coarse adjustment first, and then sharpen the image with the help of fine focusing device. Use the fine focus only when you are not satisfied with the sharpness after a careful manipulation by the coarse adjustment. Both these adjustments have limits beyond which these get either jammed or derailed from the track; these are dangerous situations and should be avoided.

Vibration Planes of the Polars

A point to check at the start of each session of the microscopic examination is the orientation of planes of polarization of the upper and lower polars. As mentioned in the previous section these should be N-S for the lower polar and the E-W for the upper polar. The manual should be consulted since a few manufactures set these planes in the opposite manner. Absorbance of light by tourmaline and biotite are very convenient tests for the lower polar. The absorbance behaviour of tourmaline is just opposite to that of biotite (Fig. 5.8A). This can be easily cross checked by rotating the microscope stage by 90° (Fig. 5.8B) and noting that when biotite grain is brightest, tourmaline grain is darkest, and also *vice versa*. Pull out the upper polar from the path of light and place

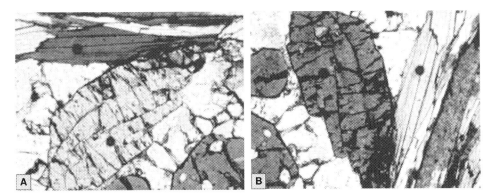

Fig. 5.8 Cross checking of alignment of polars in a polarizing microscope with the help of biotite and tourmaline. Figure B is rotated with respect to figure A by 90°.

a section containing biotite with its cleavages parallel to the E-W crosswire. It should show least absorption; if not then a slight rotation of the lower polar in either direction will bring the mineral to its brightest position. This position be marked and fixed as zero position of the lower polar. Now, take the mineral section out and with no object on the stage pull back the upper polar in the path of light. The eye now experiences complete darkness since the plane of polarization of the upper polar being E-W, *i.e.*, normal to that of the lower polar, will block the light coming from below. If the complete darkness is not seen then, you need to manipulate the upper polar either way to bring about total darkness. The upper polar then should be fixed in this position and should not be disturbed. Should there be a need to change the settings of the polars, remember to put them back to these zero positions before putting the equipment back to the cabinet. The zero position with both polars in the path of light is referred to as the *crossed nicols or polars* position of the microscope. If the small manipulations do not yield total darkness then it would mean that one or both polars have been worn out through the heat of long

The Polarizing Microscope

usage hours and a replacement is the best answer. Do not waste your time using such microscopes.

Illumination

Illumination is an important aspect of microscopy as we have mentioned above that it contributes towards resolving power of the lenses. In many models the illumination settings are done and fixed at the factory level. In such cases only a trained technician should be asked to correct any illumination related problem. On the other hand, many microscopes, especially the research models, allow user to influence illumination. The first step is close the condenser diaphragm to the maximum extent possible so that an image of a pinhole is visible. Then by adjusting the height of the condenser assembly the image of this iris diaphragm can be sharpened. This exercise also permits us to know whether the illumination system is centered with reference to the microscope axis and the centre of the crosswires. If the image of the iris diaphragm is not perfectly centered, then you should operate the two centering screws, along x and y directions, attached to the assembly of the condenser to bring the image to the centre. Now, slowly enlarge the iris diaphragm so that it just fills the entire field of view. At this juncture the entire view will be uniformly illuminated.

Objective Centring

The next step is to regularly check the centering of the microscope objectives. Since the need to do this step arises frequently, we must understand first the underlying concept of this step. We already know that there is an axis of the microscope. The principal axis of individual lens systems must coincide with this axis to a great accuracy. The first was the centering of the illumination that we have already done above. There is then, objective, Bertrand lens, Huygens ocular, and not to be forgotten the microscope stage. The centering of the microscope stage is tedious and requires expertise that is beyond expectation for an average user; for this call the service engineer of your institution. The ocular setting is normally not disturbed; if it is needed then best is to call the expert to restore the centering. *Objective centering* is often needed as we frequently change objectives during the course of single session of microscope study. Quite expectedly, all users do not handle change of objectives very delicately, though they should do so!

How do we notice that the objective requires centering? For this, note a grain, preferably a minute grain, at the intersection of the crosswires. Fig. 5.9A shows a large grain placed in the field of view approximately so that its body centre coincide with the intersection of the crosswires where as a chance there is a small rounded inclusion of another grain. Now, if we rotate the stage in either direction, and if the equipment is perfectly centered then the inclusion, being coincident with the centre, will remain unaffected in position. If it is not well-centered, *i.e.*, there is an *eccentricity*, then the mineral and the inclusion would move as shown in Fig. 5.9B; depending upon the

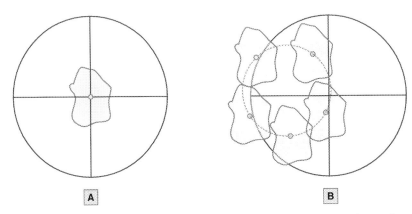

Fig. 5.9 Illustration of the procedure for centring objectives. **A.** Objective well-centred if the grains remain stationary during the rotation of the microscope stage. **B.** Off-centred or eccentric objective since the image rotate changing position through the field of view.

degree of eccentricity it may even move out of the field of view meaning thereby that it would not be visible for observation for part of the rotation.

It happens so because the objective axis and the stage axis are not coincident. Most makers of microscope provide *centering screws* that provide small X-Y motion. Use them very carefully to bring back the centering positions of the objective and the stage together. It is better to make this centering procedure as more than one stage process so that after every stage a check is made about the reduction in the degree of eccentricity. This way you would avoid tilting the centering in the opposite direction! A practical suggestion is that the microscope stage is rotated till the grain moves away to a maximum position. At this point use X-Y screws to bring the grain half way towards the centre, then manually slide the thin section so that the grain is at the centre of the field of view. Now, repeat this step till the grain remains in the centre.

A last note is necessary about the high power objective which would require use of oil between the object and the objective. Most of the oils tend to get oxidized in air; hence, it should be carefully washed from the objective lens immediately after its use is over. This way its oxidation would be prevented, no hard crust would be formed over the lens and the damage would be prevented.

Summary

A polarizing microscope is essentially a compound microscope where three assemblies of lenses constitute the main optical system. The first lens assembly is called condenser; it is placed below the microscope stage with a function to focus the light beam on the object placed on the stage. The image formation procedure involves two real and one virtual image to form. The objective forms the first real image, and it is inverted. The field lens of the ocular picks up this image and forms an inverted real image of this image. The eye piece forms a highly enlarged virtual image slightly below the stage.

The first major change in the design of a compound microscope for conversion to a polarizing microscope is the provision of a polarizing device called polarizer. The polarizer is placed below the microscope stage. The microscope stage is of circular shape and has graduation marks from 0° to 360°. With the help of a vernier the amount of rotation is noted. Another polarizing device is provided above the objective that is initially oriented in such a way that the plane of vibration of light coming out of this polar is normal to that coming from the lower polar. Between the upper polar and the objective a slot is provided in the microscope tube for the use of accessories. A high focal length system, called the Bertrand lens is situated between the upper polar and the eye piece; there is provision to put the Bertrand lens in the path of light when desired.

It is important for a microscope user to treat his microscope as delicate equipment. He should take good care of the microscope; regular cleaning especially of the oil immersion objective is essential. Objective eccentricity occurs frequently even after careful use. It can be removed through the use of two screws provided by the manufacturer.

Important Terms

Analyzer 77

Axis of the microscope 75

Bertrand lens 76

Centering screws 80

Compound microscope 64

Condenser 64

Crossed nicols 78

Eccentricity 80

Eye lens 66

Field lens 66

Huygens' ocular 85

Limit of resolution 73

Lower Nicol 71

Lower Polar 71

Numerical aperture 73

Objective centering 79

Object stage 77

Petrological microscope 77

Polarizer 76

Upper polarizer 77

Questions

1. What is the basis on which a compound microscope is so named?
2. Describe the process of image formation of an object in a compound microscope with and without ocular.
3. Explain why in photomicrography, the ocular is not placed in the path of the camera.
4. What is meant by magnification of a compound microscope?
5. What is 'least distinct of vision'?
6. Derive an expression for limit of resolution.
7. What is numerical aperture of an objective?
8. What is the purpose of using oil immersion objective?
9. Describe the characteristic features of Huygens ocular.
10. How the alignment of the plane of vibration of a polar determined?
11. Describe the procedure for eliminating eccentricity of the objective.

Suggestion for Further Reading

Bloss, F.D., 1961, *An introduction to the methods of optical crystallography*, Holt, Rinehart & Winston, New York.

Hallimond, A.F., 1972, *The polarizing microscope*, Vickers.

Johannsen, A., 1962, *Manual of petrographic Methods,* McGraw-Hill, New York.

■■■

6

Microscopic Examination of Minerals I: Orthoscopic Condition

Learning Objectives

- Observation bookkeeping
- Habit and cleavages
- Relief and pleochroism
- Polarization colors and their cause

Prologue

A crystal's optical properties depend on its structure, composition, and bonding. Indeed they provide clues as to the nature of an unknown structure. W.L. Bragg's (1924) explanation of the relation between a crystal's refractive indices and its structure requires first a closer look at the nature of light. Maxwell's equations indicated that light wave represented a time-varying electric field E inestimably associated with similarly varying magnetic field B at 90° to it. The vibration directions alluded to earlier........represented this time-varying field E.

— F. Donald Bloss

(Courtesy: P. Singh)

A photomicrograph of kink mica flakes from a metapelite from Ladakh.

Chapter's Outline

- Examination in ordinary light
- Examination in plane polarized light
- Examination with crossed polars

Introduction

A complete description of optical properties of a mineral involves several steps, but ordinarily useful information, sufficient for routine petrographic work may be obtained by observing the mineral under *orthoscopic condition*. This refers to a set up of the microscope where the object receives a beam of light incident normal to the polished surface of the grain. Normally this is easily achieved except that you may ensure that the sub-stage condenser (Chapter 5, page 65) is out of the path of light if the mechanics of the microscope allows so. Table 6.1 is a suggestion to new users for recording the systematic description of the optical properties of a mineral. You will notice that at the top of the table there are two circles where you will draw sketches of the view of the grain. Although it is given at the top, but this is one part that is to be done after the thin section has been studied very well since the sketch should be as close to the best possible view as you can find it. For the next few rows you need to consult the book and the instructor. As a beginner, take thin sections of known minerals. Nature of the phase refers to its molecular formula as well as the nature of variation in composition, *i.e.*, pure phase or solution. In the latter case we should be aware whether it is binary solution, ternary or multicomponent. This knowledge is helpful because optical properties are manifestation of internal structure and composition. Likewise, Z value, number of formula units in the unit cell of the mineral, also indicates its structure. Noting down the value(s) of refractive index of the mineral at the beginning of systematic examination immediately allows us to know whether the mineral is isotropic, uniaxial, or biaxial; we also get an idea of the relief and birefringence. Hence, these notings from a book, including the Part II of the present book, have given an idea of what to expect. You may note that there is a three fold division of record, *viz.*, *ordinary light*, *plane polarized light*, and *crossed polars*. For ordinary light the lower polar, situated below the stage, is also taken out of the path of light. In many microscopes it is not possible to do so; for others it may not be advisable to take out. Hence, the first two categories of observations are taken with lower polarizer in without much loss of data.

Examination in Ordinary Light

Habit

Habit of a mineral is the common manner in which grains of the mineral occur, though it is no guarantee that the mineral will always occur in that manner. Some minerals have one habit, but many minerals may occur in more than one manner; in the latter case a mineral may show more than one habit, and in such cases it is possible that different habits are results of different conditions of crystallization attending to the process. In such events, the habit becomes an important property. Even otherwise, a habit, quite often, guides us to the identification of the mineral. The following sets comprise the term *habit* of minerals:

Table 6.1 Systematic description of optical properties

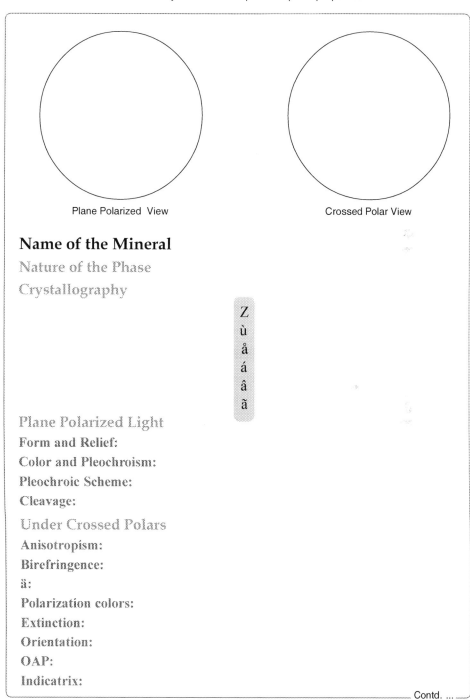

Plane Polarized View   Crossed Polar View

Name of the Mineral
Nature of the Phase
Crystallography

Z
ù
å
á
â
ã

Plane Polarized Light
Form and Relief:
Color and Pleochroism:
Pleochroic Scheme:
Cleavage:
Under Crossed Polars
Anisotropism:
Birefringence:
ä:
Polarization colors:
Extinction:
Orientation:
OAP:
Indicatrix:

Contd. ...

$2V_a$:
Sign of Elongation:
Dispersion:
Twinning:
Zoning:
Alteration
Distinguishing Characters
Mineral Association

(a) State of aggregation,
(b) Morphology (includes shape), and
(c) Size

(a) **State of Aggregation:** Some minerals tend to occur as individual grains dispersed in the rock as illustrated in Figs. 6.1 (A-D). On the other hand, there are minerals that

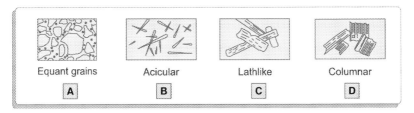

Fig. 6.1 Illustration of habits-I, forms of single grains.

tend to occur in clusters, *e.g.*, a group of 2-4 grains clustered together, and such a pattern is repeated throughout the rock; in some cases clusters may form as a group of crystals radiating out form a point. Figs. 6.2 (A-J) gives some common types of aggregations.

(b) **Morphology:** If a grain is well-bounded by straight lines we call such a grain as *euhedral*, but term *subhedral* is used when a grain presents a few straight edges but rest of the grain shows fractured or irregular outline. This property is dependent upon the crystal system to which the mineral belongs, but always bear in mind that a thin section presents a sort of 2-dimensional view. Hence, the crystal outline seen may be an artifact of the manner in which the grain got cut. Fig. 6.3 presents some euhedral habits along with Miller indices of the traces of the crystal faces that intersect the plane of the thin section written alongside. When a section is cut parallel to a particular face of the

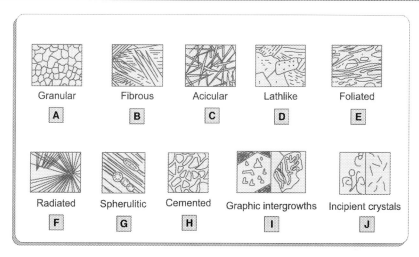

Fig. 6.2 Illaustration of habits-II, forms of aggregates.

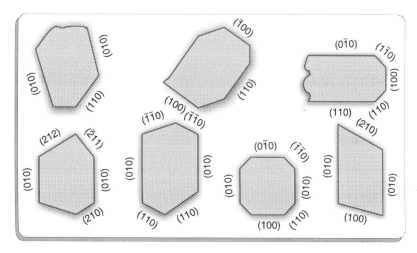

Fig. 6.3 Sections of euhedral and subhedral grains of some common minerals.

crystal then the thin section is termed as *oriented section*. Identification of such grains makes determination of optical properties easy and exact. Some typical examples are tabulated in Table 6.2. When a grain presents no identifiable edges, but shows cleavages, then also we may make a good guess as to how the section has been cut since the Miller indices of the cleavage would be known otherwise. Finally, a grain is termed *anhedral* when no straight edge of the grain boundary is seen. Table 6.3 shows how common minerals have habits that are, in some cases, sufficient as first intelligent guess.

Table 6.2 Minerals found in euhedral grains

Isometric	Tetragonal	Hexagonal	Orthorhombic	Monocline	Triclinic
Perovskite	Rutile	Apatite	Olivine	Staurolite	Plagioclase
Leucite	Melilite	Chabazite	Monticellite	Monazite	
Sodalite	Vesuvianite			Sphene	
Garnet	Zircon			Epidote	

Table 6.3 Minerals with commonly associated forms

Equant	Acicular	Lath-shaped	Columnar	Granular	Fibrous	Foliated
Quartz	Rutile	Aragonite	Quartz	Quartz	Gypsum	Muscovite
Rutile	Sillimanite	Barite	Corundum	Carbonates	Wollastonite	Biotite
Apatite	Tourmaline	Gypsum	Orthoclase	Barites &	Anthophyllite	Stilpnomaline
Lawsonite	Stilbite	Tourmaline	Sanidine	Most	Tremolite	Phlogopite
Sodalite	Natrolite	Epidote	Microcline	Sulfates	Riebickite	Chlorite
Olivine		Piedmontite	Plagioclase	Olivines	Sillimanite	Chloritoid
Garnet		Prehnite	Nephaline	Epidote	Prehnite	Anthophyllite
Zircon		Pyrophyllite	Wollastonite		Antigorite	Talc
Topaz		Kyanite	Glaucophane		Chrosotile	
Andalusite			Beryl			
Allanite			Scapolite			
Cordierite			Vesuvianite			
Sphene			Topaz			
Glaucophane			Kyanite			
Analcime			Clinozoistite			
			Barite			
			Micas			

(*c*) **Grain Size:** Many minerals tend to occur in definite sizes. For example, clay minerals tend to be micron size, whereas garnets may be of millimeter, or even centimeter scale. Here, again a grain may be cut in such a manner that one may not get a precise idea of how large a grain is. If the thin section has more than one grain then for a quick determination, always look for the largest grain for measurement. Now a days such measurements, even statistical averages etc., are done with a click of the mouse with the help of an attachment called *Image Analyzer*[1] (Chapter10, page YY). However, a

---

1. Considering the cost of image analyzer and its accessories, it may be remarked that a polarizing microscope becomes an attachment!

quick, and certainly cheaper, method is to use a *micrometer* (Fig. 6.4), which is placed below the eye lens in the ocular (Chapter 5, page 80). The micrometer, as shown in the figure is a round piece of glass of about 1mm thickness upon which a scale consisting of fine lines are engraved. About a centimeter is covered by 20 or 50 such lines equidistant apart. The location of this within the ocular is at the focal plane so that when viewed a clear image is seen. In most cases, this coincides approximately with the location of the crosswires so that both are viewed simultaneously, although, it is not necessary. Before the measurements are taken, it is essential that the micrometer be calibrated for each combination of objective and eye piece. This is done by placing a pre-calibrated grain on the stage, and, then for a given combination of objective and ocular the micrometer is easily calibrated. A table may be prepared of this calibration with different objective-ocular combinations and suitably placed in the laboratory for view. This will avoid repeated calibrations and would promote rapid size determinations. The common practice for photographic illustration has been to put a small transparent line of about 1mm long on the stage on top of the section. Currently available good microscopes with photographic attachments have automatic provision for showing the scale.

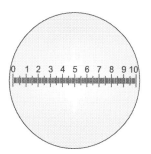

Fig. 6.4 A micrometer with graduation.

Cleavages

You already know that *cleavages* are planes of weakness that arise from the differential bonding of ions or atoms within the unit cell; therefore, the presence of a cleavage set is a characteristic property of the mineral affording a simple and reliable means of its identification. Since cleavage is an internal property of a crystal, it is always parallel to a possible crystal form; the form, itself, is characteristic of the symmetry class in which the mineral crystallizes. The cleavage is also subjected to symmetry operations of that class. In addition, a cleavage has a quality attribute as well; that is to say that presence of a weak bond, like van der Waal bond, produces such cleavages that a mineral grain may split by merely rub of a finger over its surface! Such cleavages are termed *'excellent cleavage'*; on the other hand, in several minerals, it may require a gentle stroke of a small hammer to cause splitting along a given cleavage plane. Such a cleavage is barely visible, so it is termed *'poor cleavage'*. However, you may well be cautioned that the fact that a given crystal is formed of ionic or covalent bond does not always mean absence of cleavage, because the presence of cleavage is determined by the over all linkage pattern of ions within a unit cell and not just the nature of bonding between an ion or atom and its nearest neighbours.

In hand specimens, we recognize cleavages as smooth planes that are all pervasive all through the sample and are equally spaced. Identification of the crystal form to which a given cleavage corresponds is often a tedious process particularly in subhedral grains; therefore, for common minerals, you are advised to consult the description given in textbooks for the Miller indices of the form to which a given cleavage set belongs.

It is also a common case that the internal structures of many crystals allow for more than one set of planes of weakness in their structures resulting in more than one set of cleavages. Individually, each set will be parallel to one particular crystal form. The description of the cleavage, though, will still be guided by rules set above, *i.e.*, Miller indices of the crystal form for each cleavage set as well as the quality of the individual sets must be mentioned. Number of sets, their relative orientation (the angle between cleavages of one set with nearest cleavage of another set, Miller indices, quality of cleavages of individual sets are all characteristic property of a given mineral. A systematic description involves a complete description of cleavage present. Fig 6.5 shows some typical examples of cleavages as seen in under microscope.

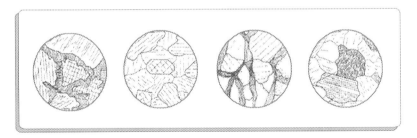

Fig. 6.5 Sketches of different cleavage sets under microscope.

Study of cleavages under microscope is beset with some typical problems of microscopy. The first is of course the thin section effect we have mentioned more than once. A thin section, as we know, tends to produce a 2-D effect; hence when a section is cut parallel to a cleavage plane, it will not show any cleavage! In minerals with more than one set of cleavage planes, only such sections that are cut normal to all sets will show all cleavages clearly, otherwise fewer sets will be visible. Therefore, in multi-grain thin sections we must browse all grains in low power objective to get an idea of the orientation of individual grains. If this procedure is embraced as a habit it will make the microscopic work easier and pleasant. When we record different sets we must ensure that multiplicity is due to different forms and not because of symmetry. For example, minerals like fluorite, diamond, calcite, pyroxenes and amphiboles show more than one set because of symmetry. Therefore, we record for fluorite one set of excellent octahedral cleavage {111}, although as much as three sets may be observed because of the presence of three fold rotoinversion axis. You may number cleavage sets for one set of excellent rhombohedral cleavage {10$\bar{1}$1} in calcite, for one set of good prismatic cleavage {210} in orthopyroxene etc. Irrespective of the reason for the presence of more than one set of cleavage, it helps to measure the angle between the two sets present. For this, bring one set of cleavages parallel to one of the crosswires (Fig. 6.6) and microscope stage reading is noted. Then the next set is brought parallel to the same crosswire, and the reading is noted again. The difference between the readings gives a value of the angle. Recalling that a true angle between two planes is measured in a plane normal to both the planes, the angle we measure between the cleavage sets should be recorded as apparent cleavage angle unless we know for sure that the angle has been measured in a plane normal to the cleavages.

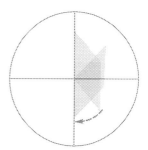

Fig. 6.6 Measurement of angle between two sets of cleavages in a mineral in thin sections.

During the preparation of a thin section, if the grain lies flat on its cleavage plane may show differential stacking of cleavage partings as shown in Fig. 6.7. Such grains show zoning of polarization colors (*see* the following section) under cross polars, a condition that must be distinguished from zoning arising due to chemical composition differences in between parts of a grain.

Fig. 6.7 A photomicrograph of a epidote under crossed polars. Note that zoning is seen by arrangement of polarization colors that is due to differences in thickness of rim and core of the grain.

Under Plane Polarized Light

Relief and its variation

We know that the quality of the appearance of the border of a grain is determined by the difference in the refractive index of the mineral and the medium surrounding the grain (Chapter 3, page 32). In an anisotropic mineral ù and å or á, â, and ã have different values; hence, the relief is to be recorded with reference to the vibration direction of the polarizer, *i.e.*, the lower polar. For most minerals a careful and slow rotation of the microscope stage ought to reveal the variation in the quality of relief. No such variation is noticed if the two refractive indices are close to each other. On the contrary, if the difference between the two refractive indices is large, particularly if the refractive index of the immersing medium lies between the two extremes of the mineral of the stage then rotation would bring strong and conspicuous changes in the relief of the mineral. Calcite is the best example as its grain boundary appears bold, disappears, and then appears negatively bold, repeating it four times in one rotation of the stage. This property is named *twinkling* in allusion to twinkling of stars.

The property of relief has been discussed in detail in Chapter 3 and Fig. 3.2 gives several examples of different qualities of relief.

Color and Pleochroism

For the thickness of a grain embedded in a thin section (30ì m) many minerals become colorless., though in a few cases these may be colored in hand specimen. However, you are cautioned that before recording the color you must rotate the microscope stage slowly for several complete rotation ensuring that the mineral does not reveal a faint color at any point of rotation. This precaution is taken because by virtue of *Pleochroism* minerals change their color as the stage is rotated. This change can occur from colorless to another color, say green, or from one color to another, like yellow to green, or from one shade of a color to another shade of the same, *e.g.*, light blue to navy blue.

The color of a mineral, under microscope, commonly results from the presence of metal ions of variable oxidation states belonging to *transition elements* in its structure[2]. In several examples, the stoichiometric composition of a grain may not have such ions, but these get substituted during crystallization under favourable conditions. Hence, a careful observation and its recording on the pleochroism of a given mineral may provide clue for a detailed crystal chemical examination. For this, the first step is to identify optic directions in the grain. For uniaxial minerals it is often easy as a basal pinacoid

---

2. Theory of color as proposed by Nassau, and widely accepted, has put forward several reasons for color of the substances taking into account the nature of bonding between atoms in the structure of substances. Presence of ions of transition elements is one of the reasons.

(square, triangular or hexagonal shaped grain) is normal to optic axis (*c*-axis), and a prismatic grain (rectangular in shape) is parallel to optic axis. However, for biaxial minerals, we may not be so lucky always, and, then, we shall require undertaking procedures outlined a little later in the book to establish the location of biaxial optic directions. Once these directions are determined, bring them one by one parallel to the N-S crosswire (vibration direction of the lower polarl). The color in this position of the grain is noted; as an example we say for rutile ù = yellow, å = reddish brown. The statement, known as *Pleochroic Scheme*, should form an integral part of the systematic description of optical properties of a mineral. The *Absorption Formula* describes which optic direction, amongst ù, å, or á, â, ã, absorbs more, again an indication of differential number of ions of transition elements present along that direction. In record we shall note, say for rutile, absorption formula: å > ù.

Crossed Polars

The polarizing microscope can be brought under *crossed polar condition* by placing the upper polar in the path of light (Chapter 5, page YY). Please recall that in this condition, the plane of polarization (also called vibration plane) of the lower polars (polarizer) is normal to that of the upper polars (analyzer). It has been already mentioned that in most student microscopes this is fixed by design at the factory level; but in many cases a provision is made to rotate the vibration planes of one or both polars. Therefore, you were already advised to ensure that these vibration planes are normal to each other before beginning the optical examination (Chapter 5, page YY). When the microscope optics are in crossed polar position, the ray of light leaving the polarizer will be vibrating parallel to the vertical (N-S) crosswire, and will be prevented from entering the analyzer since its vibration plane is along the horizontal (E-W) crosswire when no mineral is placed on the microscope stage. The field of view of the microscope when seen through the ocular in such a condition will appear dark as through the illumination of the microscope has been switched off.

The above paragraph merely re-emphasizes a good working habit that has already been stressed earlier. At the end of the last observation in the plane polarized light, you are not required to take the thin section out to test the suitability of the equipment, since this you have already tested before taking any observation. Hence, once, observation of pleochroism is over you may cross the polars and begin observing the following.

Anisotropism

This is the first and easiest property under crossed polars to be determined. An isotropic mineral becomes dark (the optical mineralogy term is *extinct*) as soon as polars are crossed. In such an event rotation of the microscope stage does not materially change the situation as the grain will always remain dark or extinct. On the other hand, anisotropic minerals invariably show the grain bright except under certain optical conditions to be described below.

The reason for this phenomenon lies in the nature of isotropic wave surface, which is a sphere, and so allows the plane polarized light to pass through it unaffected (Chapter 4, page YY). Hence, the light that was vibrating along N-S continues to do so, and, therefore, is refused admission through the analyzer. But when an anisotropic mineral is placed on the stage, it has its own two vibration directions, which in general, are not expected to coincide with those of the polars, *i.e.*, neither E-W nor N-S. Therefore, the entering light ray gets resolved in two components that vibrate along the two directions of the anisotropic grain. The result is that when the rays leave the mineral there are two rays both plane polarized[3] vibrating normal to each other (Fig. 6.8A). Since none of the two planes is along E-W or N-S both rays are able to enter the analyzer, and in doing so gets resolved further into four rays, *i.e.*, a set of two rays for each of two rays (Fig. 6.8B). In traditional Huygenian explanation, inside the analyzer there are four rays – two extraordinary and two ordinary. Extraordinary set vibrates in one plane and the ordinary one in a plane normal to it. Now the ordinary set is absorbed by the polar by virtue of its construction design; only the extraordinary pair emerges out vibrating along E-W direction allowing us to view the grain. Now, we are also in position to explain the four dark positions (extinction positions) during one complete rotation of the stage. When we rotate the microscope stage, effectively we are rotating the vibration planes of the mineral with respect to the vibration plane of the polars. A position comes when one of the vibration planes of the mineral comes to occupy N-S position. Now in such a case, the mineral allows light to pass through without resolving it further. Hence, this N-S vibrating light is stopped by the analyzer resulting in extinction. There are four such positions, in one complete rotation, each separated by a $90°$ readings on the microscope stage scale; it is so because there are two orthogonal vibration planes of the mineral and two for the polars causing coincidence four times in $360°$ rotation.

Let us now estimate the amplitude of the wave that is traversing the analyzer (Fig. 6.8C). Here AOA′ and POP′ are the vibration directions of the upper and lower polars respectively. VOV′ and WOW′ are the vibration direction of the mineral. Suppose VOV′ represent slow direction. If the phase difference between the two components is between $90°$ and $180°$, then the elliptical vibration within the upper polar is given by the ellipse shown in Fig. 6.8C. Then OC and OC′ are the amplitudes along the vibration

---

3. In fact, there is only one polarized ray coming out and it is elliptically polarized. However, the behaviour is elegantly and satisfactorily explained by assuming two plane polarized rays that vibrate normal to each other.

Microscopic Examination of Mineral I: Orthoscopic Condition

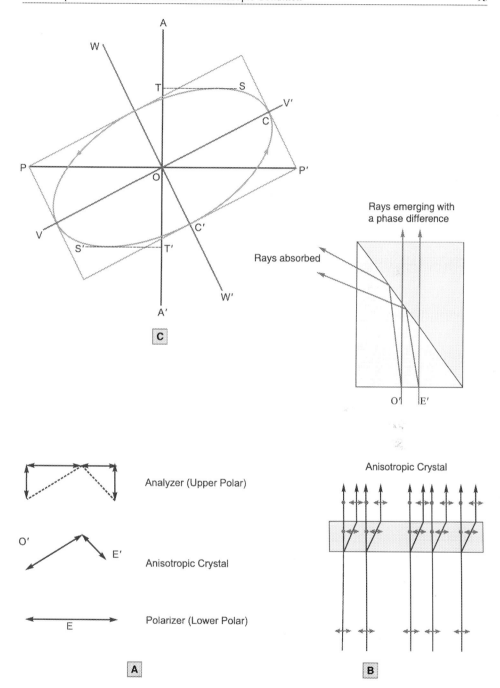

Fig. 6.8 Vibration planes in an anisotropic mineral. **A.** Traces of vibration on planes of polarizer, anisotropic crystal, and analyzer are shown in a general case. **B.** A perspective view passage of polarized light rays as it undergoes double refraction through the crystal and the analyzer.

direction of the minerals. Now, if we draw lines S-T and S′-T′ such that these are normal to AOA′ and the ellipse, then OT and OT′ are the two amplitudes of the two components used by the analyzer.

Polarization Colors

Polarization color of minerals is a property that is a direct result of the optics of petrological microscope seen with crossed polars. To appreciate the concept underlying the phenomenon of polarization colors let us begin with an illumination that is monochromatic with a wavelength λ. It is plane polarized as E as shown in Fig. 6.8; its fate through the mineral and rest of the microscope through ocular has already been described in the section above. We shall now analyze the relationship between E'' and E''', *i.e.*, the two components of the monochromatic light ray that finally emerge from the analyzer. As shown in Fig. 6.8C, OT and OT′ represent the two amplitudes.

For a grain of thickness d the slow ray will obviously require longer time to pass than that taken by the fast ray. Suppose the fast ray with a refractive index of α' takes t_1 time; further the values respectively for the slow ray are γ', and t_2. Then we have:

Velocity for the fast ray $= (d/t_1) = (c/\alpha')$; and
Velocity for the slow ray $= (d/t_2) = (c/\gamma')$; hence
$t_1 = [(\alpha'* d)/c]$; and $t_2 = [(\gamma'* d)/c]$

To find out the amount of time by which the slow ray will be behind the fast ray we may take a difference between these two times; hence,

$$t_2 - t_1 = [\gamma' - \alpha']* [d/c]$$

Therefore, the above equation tells us that the amount of time by which the slow ray will emerge later than the fast ray, alternatively speaking, the *retardation* between the two rays is governed by (*a*) $[\gamma' - \alpha']$, the difference between the two apparent refractive indices of the grain in the plane of the thin section, and (*b*) the d, the thickness of the section. The value of $[\gamma' - \alpha']$, a number, is also called its *relative birefringence*. Let us multiply the retardation $(t_2 - t_1)$ by c, the velocity of light to give us the linear distance $R_{m\mu} = [\gamma' - \alpha']* d_{m\mu}$. The subscript m$\mu$ denotes the unit micromillimeter. To get the retardation in degrees, $R_{m\mu}$ is multiplied by $360/\lambda$, and, when expressed in radians it is multiplied by $2\varpi/\lambda$. Please note that in the treatment given above we have used the term relative retardation and relative birefringence that refers to these values within the plane of the thin section. In general, this section is not the principal section of the uniaxial mineral or the optic axial plane of the biaxial mineral (Chapter 4, page 59). The true birefringence is always measured within these sections, *i.e.*, *true birefringence* is given by å–ù for uniaxial minerals, and ã–á for biaxial minerals.

According to the above discussion it is obvious that for a grain of a given mineral in a thin section [ã´–á´] value is fixed; hence the retardation becomes a function of its thickness. If we prepare a thin section of a mineral, say quartz, in a wedge like manner as shown in Fig. 6.9 then a ray of light entering at a given point through this wedge will suffer a retardation whose value will be different from that ray which enters the wedge

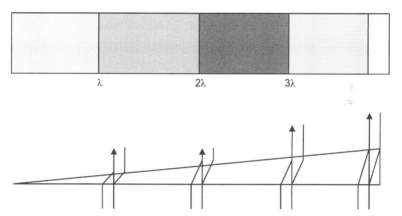

Fig. 6.9 Quartz wedge. It illustrates the control of thickness of the grain on the intensity of a monochromatic beam.

through another point. Fig 6.9 illustrates four rays that enters the wedge in such a manner that the relative retardation caused by each is an integral multiple of ë. When the retardation is ë or an integral multiple of ë then, the two emerging rays will suffer complete destruction and darkness will result. However, between the two darkness positions the maximum resultant amplitude will occur when the value of the relative retardation will be 1/2ë or any of its odd multiple. You know that the intensity of light is proportional to its amplitude, so this is the position of maximum illumination for the wavelength.

Now, it must be obvious to you why we mentioned earlier (Chapter 9, page YY) that the thickness of the grain is kept at 0.03mm by convention. It is because the relative retardation caused by a mineral is dependent on the thickness of the grains through which the light ray has to pass. Another convention is to use sunlight, also called white light, for routine microscopic examination. The sunlight is essentially a polychromatic light (*see* Fig. 1.1, page 6) which we speak of as consisting of seven colors, *i.e.*, seven wavelengths, although at macro-level the variation is continuous. These seven wavelengths will have different bandwidths with reference to Fig. 6.9. The physical effect is that when relative retardation of a given wavelength reaches its ë value, it will alone be extinguished; other wavelengths will have some amplitude, one of them may be strong enough to dominate. So at that point the quartz wedge will show that color. Fig. 6.9 graphically shows the bandwidth of each color with relative retardation on *x*-axis. As an example we can begin at the extreme left hand corner where the wedge is of almost nil thickness, no ray is allowed; so we see a black band. With a slight increase in thickness, a phasal difference is introduced which we know will be greatest for the blue giving a grayish blue tint. Further increase in thickness brings us into a central region for all wavelengths resulting in white color, but as move into thicker regions progressively, violet and blue extinction positions are approached, which as seen in Fig. 6.10, almost coincide with the central region of yellow allowing us to view yellow for this thickness. Such a color is not an intrinsic property of the mineral but arises due to interference of light and is dependent upon birefringence and thickness. Such a color is called *polarization or interference color*. As the retardation increases because of the progressive increase in the thickness, a stage is reached when all colors, except red, become weak due to unfavourable phasal difference. This completes 1ë for all wavelengths, hence we term this color spectrum including red as the *First order of polarization colors*. Then increase in thickness results in violet color and so on finally ending again red. The red of the first order is achieved around 550mì and that of the second around 1000mì of retardation. However, again a look at Fig. 6.9 will exhibit that width of wavelengths are spacing progressively wider causing a complex interference pattern as the retardation is increasing. It results in higher order colors becoming fainter and finally a white color is achieved. In order to distinguish this white color from that arising from a low retardation, you rotate the upper polar to make its vibration direction

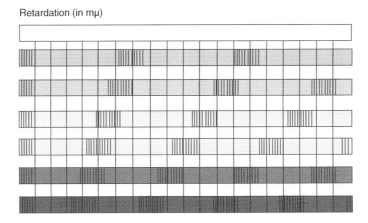

Microscopic Examination of Mineral I: Orthoscopic Condition

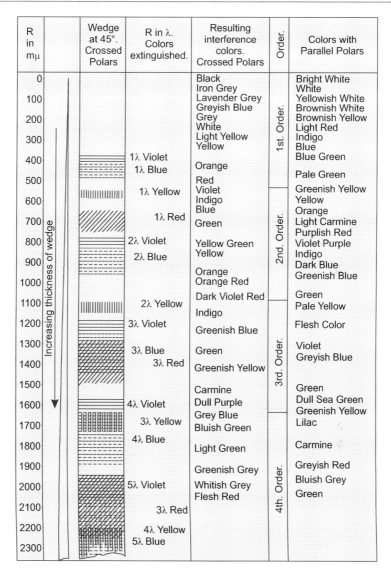

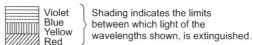

Fig. 6.10 A graphical representation of the interference color chart. A sketch of quartz wedge (Fig. 6.9) is reproduced rotated 90° in column 2 for illustrating the effect of increasing thickness (therefore retardation also) on the resulting color.

parallel to that of the lower polar. First order white yields to indigo in this condition, and green or carmine color, if the higher order white has been achieved.

It must be obvious to you that this description of interference of rays of sunlight to produce a spectrum of colors is similar to the colors produced from reflection of sun rays from a thin film of oil floating over water. It is the well known *Newton's scale of colors* that you have read in the school science class. Taking cue from this Michael Levy prepared a chart at the beginning of the last century that is known after his name and is not, unfortunately, available in its original form anymore. However, its smaller and somewhat poorer versions are available in many textbooks including the present one (Fig. 6.11 as a sketch; a flap at the end of the book). It consists of a rectangle with its Y

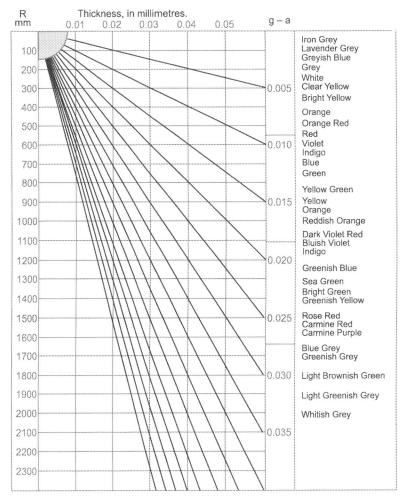

Fig. 6.11 A sketch of the Michael Levy chart given as a flap at the end of the book.

direction showing the thickness of grains and X direction indicating relative retardation in mì in the bottom and birefringence (ã á; å – ù) values written along the top. From the left hand bottom corner a number of diagonals are drawn that intersect the top and right hand lines of the rectangle where a number of mineral names are written. It has been a very popular chart, second only to the periodic table, in mineralogy class rooms and individual labs. As an example let us take a grain of quartz cut oblique to its c-axis and is of thickness 0.03mm, the standard thickness. Under cross polar it shows grey polarization color. You may independently find this out without recourse to looking into the ocular by following the name of minerals written on the top line and spotting quartz name. Then tracing back the diagonal closest to the written word quartz to the horizontal line denoting the thickness of 0.03mm we find that at the point of intersection of the diagonal and the thickness line the color is grey. Alternatively, we see polarization color of a grain as 2nd order green, then from the intersection point on the thickness line we trace diagonal to find that calcic amphibole name is written. From here on we may proceed to confirm this identification from other properties. This chart also enables us to measure *thickness of a grain* (of known mineral) in the section by the following the above-mentioned guidelines.

Anomalous Polarization Colors

Under certain conditions minerals show polarization colors that are not part of Newton's scale. These are termed as *anomalous polarization colors*. There are three main reasons for such an anomaly. First, many minerals may, under certain circumstances, incorporate ions that cause strong absorption; titanium in biotite or amphibole is an example. In many cases this substitution causes deep coloration under ordinary light and polarization colors may be masked to such an extent that one hardly notice any change in color between uncrossed and crossed polars conditions. In many examples, the grain may remain colorless or faintly colored under plane polarized light, but produce anomalous colors under crossed polars. It is caused by the dispersion of birefringence and dispersion of extinction angle (Chapter 8, page YY). Both these dispersions may affect the bandwidth of individual colors in a non-uniform manner, hence, the color scheme given in Figs. 6.10 and 6.11 will not be followed giving rise to anomalous colors.

Summary

Microscopic examination and book-keeping of observations are divided into three categories, namely, ordinary light, plane polarized light and crossed polars. For most microscopes, ordinary light and plane polarized light observations do not require separate set up. Habit of a mineral is easily recognized through recording of state of aggregation, morphology (shape) and size of grains. Next property to be recorded is the presence of traces of cleavage planes. The quality of appearance, Miller indices, number of sets of cleavages, relationship between sets are important points for the purpose of records.

Under the plane polarized light the first important property to be recorded is the relief, *i.e.*, the boldness of the outline of the grain. The relief of a mineral is controlled by the differences in refractive indices of the mineral and its surrounding material/mineral. The mineral may be colorless or show a color. When colored, the grain may show variation in color; the phenomenon is called pleochroism.

With polars crossed, the behaviour of light changes because analyzer (upper polar) allows light to pass through if it is vibrating along E-W crosswire. If the vibration plane of the light approaching the analyzer is inclined then only its component along the E-W crosswire is permitted to go through. Consequently, if the light is vibrating along N-S crosswire, the direction of the polarizer, then it is stopped and extinction of the grain occurs. In a general case, none of the vibration planes of the mineral would coincide with that of the lower polars, hence, the splitting of the rays in two parts would take place within the grain itself. The two rays would be further split in the upper polar. However, only two rays with a phase difference between them would emerge out of the upper polar. Interference of the two rays would take place; for a beam of white light, an array of colors would form. These colors are known as polarization colors. The nature of these colors is controlled by the thickness of the grain and the phase difference between the two emerging waves.

Important Terms

Absorption formula 93

Anhedral 87

Birefringence 84

Cleavages 89

Crossed polars 89

Crossed polar condition 93

Euhedral 86

Excellent cleavage 89

Habit 84

Image analyzer 88

Interference color 98

Micrometer 89

Morphology 86

Newton's scale of colors 100

Ordinary light 84

Oriented sections 87

Orthoscopic condition 84

Plane polarized light 84

Pleochroism 92

Pleochroic scheme 93

Polarization 96

Polarization color 96

Poor cleavage 89

Relative birefringence 96

Retardation 96

State of aggregation 86

Shape 86

Size 86

Subhedral 86

Twinkling 92

Transition elements 92

Thickness of a grain 101

Questions

1. What is meant by the term 'orthoscopic condition'?
2. What are the three broad categories under which systematic orthoscopic examination of a mineral is carried out?
3. Why is meant by habit of a mineral? Why it is so important?
4. How will you describe the habit of a mineral under microscope?
5. If a mineral shows two distinct grain size, then how many explanations we may put forward for test?
6. Is it possible that a particular grain of a given mineral does not show cleavage traces though the mineral possesses cleavages. Explain.
7. Elaborate the term 'twinkling' and its cause.
8. Define the term pleochroism?
9. What is the importance of pleochroism? How does one maintain its record?
10. What is the meaning of absorption formula? Give an example.
11. Why does an isotropic mineral become dark under crossed polars?
12. What is the role of vibration directions of a mineral grain placed for observation during crossed polar condition?

13. Give geometric treatment of the passage of a light ray through an anisotropic mineral during crossed polar condition.

14. What is meant by interference color? Give its theory.

15. Describe Michael Lévy Chart.

16. Explain anomalous polarization colors. Give examples.

Suggestion for Further Reading

Kerr, P.F., 1959, Optical Mineralogy, 3rd ed., McGraw-Hill Book Company, Inc.

Nesse, W.D., 2003, Introduction to Optical Mineralogy, 2nd ed., Oxford University Press.

Whalstrom, E.F., Optical Crystallography, 5th ed., Wiely, N.Y.

Wincell, A.N., 1937, Elements of Optical Mineralogy, 5th ed., John Wiley & Sons Inc., N.Y.

■■■

7

Microscopic Examination of Minerals II

Learning Objectives

- Extinction positions and extinction angle
- Crystal twins and their optical effects
- Disequilibrium during crystallization and their optical effects
- Optical anomalies
- Concept of compensation of retardation and use of compensators

Prologue

The polarizing microscope, like any other microscope, will only give the best results if the user understands and applies these (principles of optics) principles. Too often a student is given a microscope without any guidance in these matters, and he 'muddles on', regarding the instrument simply as a super hand lens with the added complication of polars.

— A. Stuart & N.H. Hartshorne

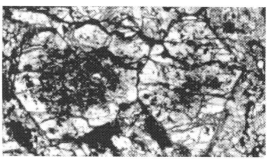

(Courtesy: P. Singh)

Garnet grains from an eclogite from Ladakh exhibiting zoning and overgrowth.

Chapter's Outline

- Extinction and extinction angles
- Twinning
- Zoning and overgrowth
- Optical anomalies
- Compensators

© The Author(s) 2023
P. K. Verma, *Optical Mineralogy*,
https://doi.org/10.1007/978-3-031-40765-9_7

Examination in Crossed Polars (Contd.)

Extinction and Extinction Angles

Much has already been written about state of darkness (= extinction position) in the preceding pages. In a single grain view the entire field of view becomes dark, but in the multi-grain view (all grains of the same mineral or of different minerals), only a particular grain, or a few grains will become extinct or dark; rest of the grains in the field of view would continue to exhibit brightness. Let us summarize the theory related to extinction already learnt so far (Chapter 6, page YY). It states that extinction of a grain of a mineral occurs when the grain is viewed under crossed polars if the section is cut:

1. parallel to a circular section of the indicatrix; or
2. parallel to an elliptical section of the indicatrix and its vibration directions are parallel to those of the polars.

The basal sections of the uniaxial minerals will remain extinct even when the stage is rotated by any amount of angle. Sections cut normal to an optic axis of a biaxial mineral will exhibit faint light owing to a fact that these axes are not directions of single ray velocity. The faint illumination will remain unchanged during the rotation of the stage.

Extinction Angle of a mineral is given by the angle between the optic direction and the nearest crystallographic direction measured in the plane that contains them. Fig. 7.1A shows a plane polarized view of a subhedral grains with good cleavages; the microscope stage is rotated till the cleavages are aligned parallel to N-S crosswire; position of the zero of the vernier is noted. In the next step, the polars are crossed, and the stage is rotated till the grain becomes extinct as shown in Fig. 7.1B; now the reading on the vernier attached to the stage is again

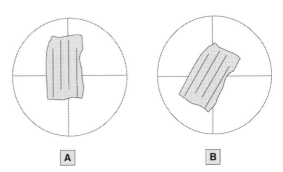

Fig. 7.1 Measurement of extinction angle. **A.** Cleavage traces brought to coincide with N-S crosswire; plane polarized view; **B.** Crossed polar view; the situation obtained after the grain is made to become extinct through rotation of stage.

noted. The difference between the two readings is the numerical value of the extinction angle. The situation is simple in the case of uniaxial minerals since the optic direction is always parallel to crystallographic direction; hence the extinction angle is zero. Sections that exhibit zero degree extinction angle are said to exhibit *parallel* or *straight* extinction (Fig. 7.2A). In cases where a section reveals two traces of cleavages or faces and the extinction direction bisects the angle between the two, then the extinction is said to be *symmetrical* (Fig. 7.2B). In other cases, we term the extinction as *oblique* or

Microscopic Examination of Minerals II

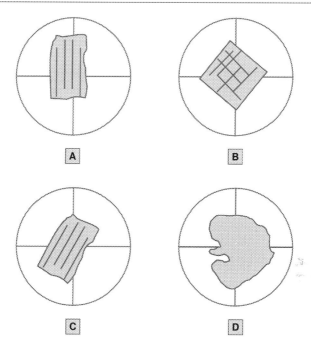

Fig. 7.2 A. Parallel or straight extinction; **B.** Symmetrical extinction; **C.** Inclined extinction; **D.** A case of minerals that does not have defined extinction angle.

inclined (Fig. 7.2C). Extinction angle measurement is meaningless in the absence of cleavage traces, or a well-defined euhedral or subhedral outline as depicted in Fig. 7.2D.

The mention of a euhedral tetragonal crystal flashes the well-known zircon crystals in our imagination. It consists of a combination of one or two prism forms terminated by pyramids. In several examples, basal pinacoid form is also associated. Most tetragonal crystals are composed of these three forms and occur as elongated grains along z-direction. Because of this common habit, it is easy to discern whether a given grain in a thin section has been cut along x-y (a square or diamond shaped outline), or y-z (x-z) (long prismatic laths) planes. But a section may also be cut in a random direction as well. The author would like you to appreciate that the appearance of cleavage traces are affected by the manner of thin sectioning, therefore, the interpretation of extinction angle should be done after due consideration of two factors:

    A. The nature of the two directions being measured; and
    B. The orientation of the plane in which the above directions are being measured with reference to x, y, z direction.

This is well-illustrated in Fig. 7.3. Fig. 7.3A is a euhedral grain with prominent prisms and pyramids. In our hypothetical example we assume that the mineral has a prismatic cleavage set {110} and also a basal pinacoid set {001} as shown by dashed lines. An outline of uniaxial indicatrix is also shown embedded within the crystal. If a thin section is made parallel to basal pinacoid, only one set of two prismatic cleavages would appear as shown in Fig. 7.3B. The indicatrix section, being normal to optic axis, is a circular section with radius equivalent to ù, and the grain will behave like an isometric crystal. On the other hand, a prismatic section, section parallel to optic axis, will show an elliptical indicatrix section with one semi-axis equivalent to ù and the other semi-axis equivalent to å (Fig. 7.3C); the relative lengths will be determined by the optic sign of the mineral. The grain will exhibit two sets of cleavages but one of these is prismatic {110} and the other is pinacoidal {001}. This section shows maximum birefringence; since, å and ù are parallel to cleavage traces, the extinction angle is zero (parallel or straight extinction). The mineral may be cut in such a manner that the cleavage sets are normal to the rectangular outline of the grain (Fig. 7.3D) this will also yield parallel extinction though the birefringence is now lower being apparent birefringence

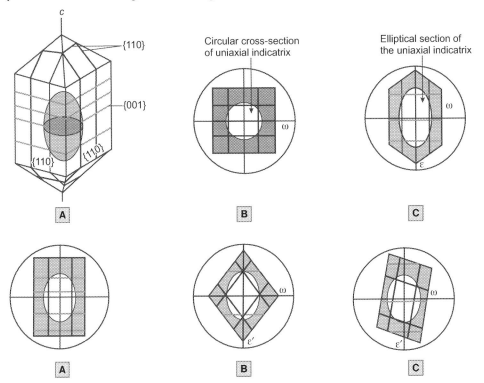

Fig. 7.3 Extinction angles situations in tetragonal crystals. **A.** A typical habit of a tetragonal crystal with prismatic and pinacoidal cleavages and location of uniaxial indicatrix; **B.** Basal pinacoid section exhibiting prismatic cleavages along with circular section of indicatrix; **C.** A section cut parallel to the optic axis. The two sets of cleavages are now consisted of prismatic and pinacoidal form; **D.** A rectangular section of the grain with cleavage sets normal to the outline; **E.** A diamond shaped cross section of the grains with cleavages so oriented to cause symmetrical extinction; **F.** A random section showing inclined extinction.

in the plane of the thin section. However, a section where outline of the grain are now in diamond shape, the extinction is symmetrical as shown in Fig. 7.3E. Finally, an inclined extinction would result in another random section (Fig. 7.3F).

For hexagonal crystals we may consider the well-known case of rhombohedral cleavages. Complete extinction is observed in the normal to optic axis section. In the optic axis parallel section, the grain may be cut perpendicular to a plane of symmetry; in this case all three cleavages are visible as shown in Fig. 7.4A. The extinction angle is zero with respect to one of the three cleavages but symmetrical for other sets. A section of the grain may be cut in other directions, where, again, all three cleavage traces are visible but in this case none of the cleavages show parallel extinction, but each will show inclined extinction (Fig. 7.4B). In the absence of the rhombohedral cleavage, extinction angle is measured with the help of basal and prismatic cleavages. Such cases are illustrated in Fig. 7.4C. The section has been cut in a random direction both prismatic and pinacoidal cleavages are easily seen pinacoidal cleavage, being parallel to ù, shows parallel extinction, but all the three prismatic cleavage traces will normally show inclined extinction. For the oriented sections in this case the behaviour is identical to the cases described above.

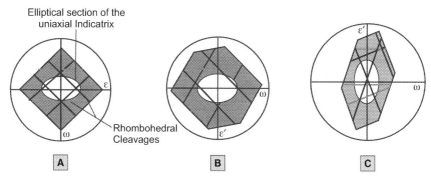

Fig. 7.4 Extinction angles situations in hexagonal crystals. **A.** A section cut parallel to the optic axis of a hexagonal grain that is characterized by the presence of rhombohedral cleavages. Section is normal to one of symmetry planes. One of the rhombohedral cleavages indicates parallel extinction while the other two are inclined; **B.** A random section of the grain where none of the cleavages shows parallel extinction; **C.** A hexagonal crystal section that has no rhombohedral cleavages but has pinacoidal and prismatic cleavage; it is a random section. In this situation pinacoidal cleavage traces show parallel extinction but prismatic cleavage traces show inclined extinction, in general.

In orthorhombic crystals, all three optic directions, á, â and ã are parallel to a, b and c. Therefore, sections cut parallel to any face that belongs to crystal zone whose axis is one of the crystallographic axes will also have zero degree of extinction angle with reference to any cleavage that is also parallel to one of the crystal directions (Fig. 7.5A). Other planes exhibit non-zero extinction angle except basal pinacoid section, which will show symmetrical extinction (Figs. 7.5B and C).

Optical Mineralogy

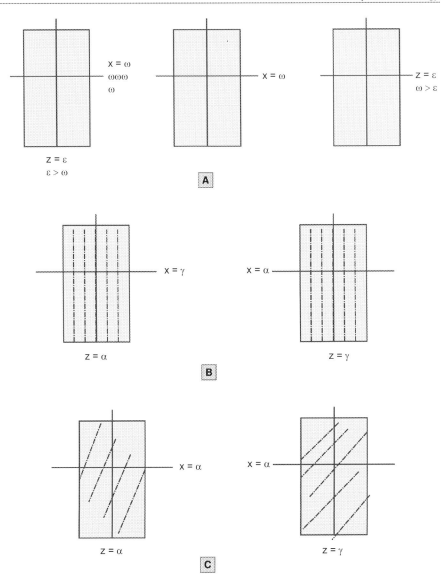

Fig. 7.5 Extinction angles situations in orthorhombic crystals. No á, â, and ã directions are given in these general cases. The optic directions coincide with crystallographic directions though in individual orthorhombic minerals the relationship varies. **A.** An orthorhombic crystal cut parallel to c-axis exhibit parallel extinction; **B.** A random section cut through an orthorhombic crystal showing inclined extinction; **C.** A basal section with symmetrical extinction.

In the monoclinic crystals one optic direction is along the *b*-axis, the other two lie in a plane parallel to (010); however, the orientation of the OAP is not necessarily parallel to (010) as the OAP is located always normal to â optic direction (Fig. 7.6A).

Hence, the [a] zone axis and [c] zone axis will have planes that would show straight extinction (Fig. 7.6B) but other sections would exhibit oblique extinction (Figs. 7.6C and D). It is interesting to note that when (100) section of a monoclinic mineral is viewed it would show parallel extinction but as we take crystal planes that are progressively more inclined to (100) the angle will rise to attain its maximum value normal to it, *i.e.*, when the plane is parallel to (010). This variation is controlled by the inclination of the section to (010), location of OAP, and the 2V.

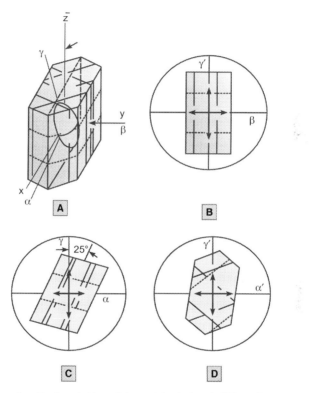

Fig. 7.6 Extinction angles situations in Monoclinic crystals. **A.** A typical Monoclinic crystal with prismatic and pinacoidal cleavages and hypothetical biaxial indicatrix located within it; **B.** A section cut parallel to (100) will have cleavage traces parallel to vibration directions; **C.** Section cut parallel to (010); it shows inclined extinction; a random section will also show inclined extinction; **D.** Random section; it shows inclined extinction.

In the triclinic system, no optic direction is parallel to any crystal axes; hence, oblique extinction would be the rule.

It may be noted that in cases where the monoclinic or triclinic minerals show chemical compositional range, the biaxial ellipsoid will rotate from one given orientation in one zone to another. Hence, for a specific composition, an optic direction may become parallel to the crystal direction making way for straight extinction where none would normally be expected. In all such cases, the value of extinction angle carefully measured would allow us to make an estimate about the chemical composition of such solid solutions.

Measurement of Extinction Angle

In a non-foliated rock, where a number of grains of a given mineral are present in a thin section, it is easy to spot euhedral to subhedral grains. The grain boundary of such grains gives an idea of the orientation of the grains. Moreover, when cleavage traces can be seen, then also one can visualize the orientation; for strongly cleaved minerals, *e.g.*, micas, the absence of cleavage traces means that the section is parallel to a cleavage plane. Large numbers of uniaxial minerals have habit of forming elongated prismatic grains (parallel to *c*-axis) with a square, diamond shaped, triangular or hexagonal basal section (depending upon the crystal system of the mineral). Similarly, habit of individual biaxial minerals also form characteristic sections. Such criteria enable us to know the orientation of the grains.

Once grains with different orientation are selected, the crystallographic direction of a given grain is brought parallel to one of the crosswires, say N-S crosswire, under plane polarized light (Fig. 7.1A). It is recommended that you view this under plane polarized light as it is more illuminated than the crossed polar position. In the case of the strong dispersion of the extinction position (no complete darkness), a monochromatic source should be used. After parallelism between the crystal directions with the crosswires is attained, the vernier reading is noted. Now the upper polar is inserted in the path of the light (crossed polar position), and the microscope stage is rotated till the nearest extinction position is attained (Fig. 7.1B). In other words, if the extinction is not achieved up to 45° rotation then one should reverse the direction of rotation. The vernier reading is again noted; the difference would give the extinction angle. For recording as an example we write $z \wedge \acute{a} = 15°$. To confirm this now reverse rotation to note the complimentary angle. More than one similar grain should be examined and an average taken provided the spread of values is small. A large spread means that apparently similar grains are not with the same orientation. In such cases, resume your selection of grains. In the succeeding chapters we shall describe other methods that are more accurate but more time demanding.

Twinning

Amongst the crystallographically controlled intergrowths, twinning is ubiquitous amongst natural occurrences. A large number of minerals show this phenomenon (Technical Text Box 7.1). Since different individuals of the aggregate have opposite

Technical Text Box 7.1

TYPES OF TWINS

When all the grains of an aggregate are of one mineral, and, occur such that each grain always bears a specific crystallographic relation with its neighbouring grains in respect to a possible face or an axis, then the aggregate is termed the twinned aggregate. Twins are easy to recognize in hand specimens by the presence of re-entrant angles at corners. Twinned crystals exhibit higher symmetry than that originally present. For example, twinned gypsum (Fig. 1A) shows an additional mirror plane along (100) which we know is not a mirror plane in gypsum (Fig. 1B). The type of twin illustrated in Fig. 1A is called a contact twin, which means that two individual grains of a mineral are in contact with each other with the contact plane being (100) and is termed twin plane. The statement that in gypsum twinned crystals occur with twin plane (100) is called its twin law. In case of a few minerals there are more than two individuals that are repeated with their contact plane being (100), or any other plane; such aggregates are termed polysynthetic or repeated twins (Fig. 1C). When, in repeated twins, twin planes are not parallel but inclined, then the aggregate may acquire different shapes as shown in Fig. 1D, and may be termed cyclic twins. In many cases, two individual grains of the mineral may penetrate each other instead of being in contact. This type of twinning is known as penetration twins (Figs. 1E and F).

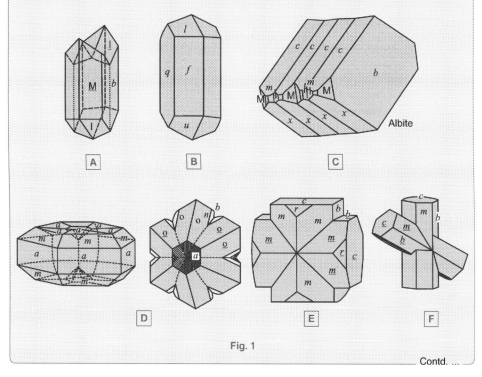

Fig. 1

Contd. ...

An aggregate may be termed as complex twin (as opposed to simple twins) when different individuals within the aggregate have more than one crystallographic relationships.

The two individuals or two adjacent individuals in any twinned aggregate have different orientation of their crystal structure with respect to each other. In most cases, the relative orientation is such as if relative rotation of 180° has occurred in one individual with reference to the other. This implies that for an anisotropic mineral the optic orientation of the two adjacent individuals are also relatively rotated. Consequently, the two individuals will become extinct at different setting of the microscope stage, although under plane polarized light the two individuals appear as a single grain. This is illustrated in Figs. 2A and B.

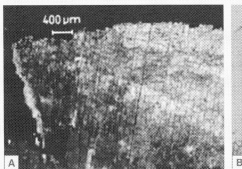
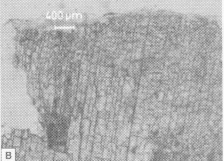

(Courtesy: Department of Geology Collection, DU)

Fig. 2 Photomicrographs of a twinned apatite grain. A. Plane Polarized Light; B. Crossed Polars.

orientation of internal structure, hence of indicatrix as well, it is easy to identify a twinned aggregate under microscope.

A twin appears as single grain under plane polarized light but appears to be an aggregate of two or more grains under crossed polars. Therefore, the two individuals of a twinned grain will show different polarization colors. As one rotates the stage, it will never become extinct as a single grain; rather the twin individuals will become extinct at different stages of the rotation as a result of difference in the orientation of their respective indicatrix. Identification of twins is often a confirmatory test for many minerals.

Let us look at Fig. 7.7. We notice that in some cases twins are simple, *i.e.*, the grains show a straight edge separating the two individuals. These are *simple twins* (Fig. 7.7A). On the other hand, in other cases the individuals are multiple but the twin planes are parallel; these are called *polysynthetic or lamellar twins*[1] (Fig. 7.7B). *Complex twins* are those grains that exhibit presence of more than one twin law; in such cases lamellae crossed each other (Fig. 7.7C) often terminating one or the other.[2]

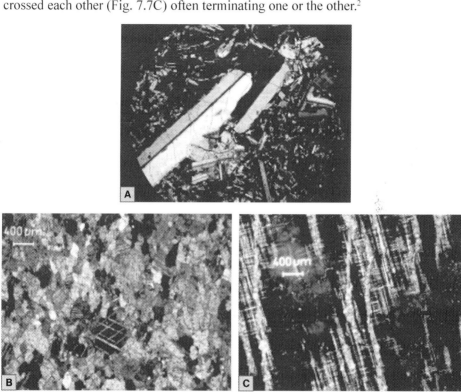

Fig. 7.7 Photomicrographs of typical twins. **A.** A simple contact twin called albite twinning in plagioclase; **B.** A poly-synthetic twin in calcite; **C.** Cross-hatched twinning arises from the presence of two twin laws in microcline; albite law with twin plane as (010) and pericline law with twin axis parallel to (010).

Not all minerals form twins; several minerals form twins under severe conditions. On the other hand, many common minerals occur invariably as twins. But you must remember that while a re-entrant angle helps in identifying a twin crystal in hand specimen, if you want to study a twinned crystal under microscope, various individuals of a twin must have different orientation of their indicatrices in order that these are

---

1. Lamellar structure may also be produced by the process of exsolution. Many a time it is easy to distinguish between a twin lamellae and an exsolution lamellae. Be cautioned that in some cases it may be difficult.
2. In many brittle minerals, *e.g.*, quartz, internal granulation occurs during deformation that also causes variable extinction in different parts of the grain. It is termed undulose extinction. It is differentiated from twinning by the absence of a well-defined twin plane.

identified as twins under microscope in crossed polar conditions. In some minerals like quartz, staurolite, etc. such a situation does not happen; hence, in such cases we must depend on physical examination rather than optical examination when searching for twinning.

Zoning and Overgrowth

We all know that crystallization of a grain is a rate dependent process involving several factors. Often, during the course of crystallization favourable factors are not sustained for some duration giving rise to imperfection in crystal structure and chemical composition. In crystallization from a melt (igneous rock formation or hydrothermal precipitation), such imperfections are not uncommon. During the crystal growth under metamorphic conditions such imperfections may, also, arise, though normally a metamorphic process is considered an equilibrium attainment process, thus, should lead to removal of any imperfection in the grain. One consequence of imperfection is the presence of twinned crystals described above; the other effect is a variation in the chemical composition of the grain if the crystal chemical consideration of the mineral allows such a variation. The compositional variation can be continuous from the centre of the grain up to its boundary, or it may form discrete zones around (concentric) the centre of the grain. The outcome is the change in refractive index, color, and birefringence. A distinct change in color can be easily seen in plane polarized light (Fig. 7.8A), and that of birefringence is observed under crossed polars position (Fig. 7.8B). Such a situation, (*i.e.*, Figs. 7.2A and B) is called *zoning*. We have noted in the earlier that chemical composition controls the optic orientation (Chapter 4, page YY); hence, for the non-isometric minerals extinction will be affected in zoned grains. For zoned grains showing continuous compositional variation a wave of extinction will appear from the centre outward. In uniaxial minerals the extinction, as we learnt earlier, is often straight, hence, no variation in the extinction angle is possible unless it is a random section. Similarly, in orthorhombic crystals, sections cut parallel to zones containing crystal axes show parallel extinction, hence, chemical zoning will not affect extinction position from one part of the grain to the other. But sections cut across the axes will reveal different zones on the basis of variable extinction. However, variation in birefringence will be apparent in the above-mentioned cases. For orthorhombic crystal that are sufficiently coarse grained and form broad zones, an estimation of 2V will also reveal the presence of zoning. When the zoning in grains is of discrete bands, then, each band will exhibit a distinct extinction angle (Fig. 7.8B), which can aid us in estimating the composition. In certain minerals, a particular face is prone to give preference for a given ion; extinction in this case may give rise to 'hourglass' structure as shown in Fig. 7.8C).

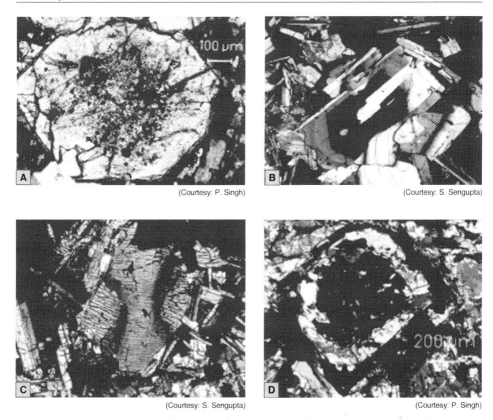

Fig. 7.8 Photomicrographs of Zoning and Overgrowth. **A.** Variation in optical properties due to compositional changes within a grain as seen under plane polarized light; **B.** Variation in optical properties due to compositional changes within a grain as seen under cross polars; **C.** Zoning leading to 'hourglass structure'; **D.** Overgrowth in garnet grains pointing to more than one phase of metamorphism.

Overgrowth is different from zoning in the sense that a distinct break in crystallization is a necessary condition in this case. In clastic sedimentary rocks, like sandstone, it is commonly a result of diagenesis where an original clastic grain of quartz acted as a seed around which new quartz was deposited. In metamorphic rocks that got metamorphosed again at a later date, *i.e.*, polymetamorphism, overgrowth of garnet is common (Fig. 7.8D). In igneous rocks, overgrowths are observed though not so frequently[3].

---

3. Most authors restrict the term 'overgrowth' for cases where the seed grain and overgrown grain are of the same minerals. When both are different we use to term 'corona' structure. Coronas are common in igneous and metamorphic rocks during their cooling phase.

General Text Box 7.1

OTHER TOOLS TO STUDY ZONING AND OVERGROWTH

For studying zoning and overgrowth, a polarizing microscope is a good starting tool. However, we would like to mention that during growth of a crystal under all four major natural processes (igneous, metamorphic, sedimentary, and hydrothermal) disequilibrium may prevail though at the scale of optical microscopy it may not visible. For example, during carbonate deposition in a sedimentary basin carbonate grains may be zoned though not noticed by the techniques described in this book. Carbonate mineralogists and petrologists take the help of cathodoluminscence spectrophotometer (CL spectrometer) for studying zoning in carbonates. It is an accessory to the polarizing microscope and uses cathode rays instead of sunlight. Similarly, in many instances, optically uniform grains reveal conspicuous zoning of element distribution when checked through Electron Probe Microanalyzer (EPMA). Currently, a large number of workers are engaged in illuminating distribution of trace elements through Ion probe microanalyazer, and Laser Ablation Inductively Coupled Plasma Mass Spectrometer (LA-ICP-Mass).

Optical Anomalies

A known mineral, say, garnet, has optical properties by virtue of which it can be identified easily. However, in a given instance, grains of garnet in a rock may undergo internal granulation followed by annealing. Such grains would appear morphologically like any garnet grain, but the granulation, when severe, could cause distortion of its spherical indicatrix into uniaxial indicatrix. This results in anisotropism of the grain that is an *optical anomaly* since garnet is isotropic. Similarly, uniaxial minerals may appear biaxial after suffering strain. On the other hand, it is possible that a plagioclase twin composed of multiple fine twin lamellae, may appear in plane polarized light like a monoclinic mineral of higher symmetry.

Such anomalous behaviour could lead to misidentification of the grains. If you suspect anomalous behaviour, then the grain should be studied under a different set of conditions for the illumination. The simplest is to insert condenser in the path of the light, or change the size of the aperture of the iris diaphragm. In addition, we must take care that no external light of significance is directly falling on the thin section (switch off the lab light or shade the stage). This exercise allows us to observe the internal reflections from the inclined surface of the granulated grain that gives out the true nature of the mineral.

Properties Requiring the Use of Compensators

In the preceding chapters, it has been mentioned that within the plane of a thin section, two optic directions are present irrespective of the fact that it is an oriented or random section, or whether it is a section of a uniaxial or biaxial mineral. One of these directions would allow light to vibrate faster than the other; if the grain is elongated, (*e.g.*, prismatic, bladed, acicular, or fibrous habit) along the fast direction we term it *length fast crystal*, otherwise, it is called *length slow crystal*. Many authors use terms 'positive elongation' (syn. length slow crystal) and 'negative elongation' (syn. negative elongation) particularly in relation to uniaxial minerals. Determination of length fast and length slow characters reveals the optic directions: ù and å for uniaxial, and á, â, ã for biaxial minerals. We know that optic orientation of a uniaxial mineral has a constant feature in that z-direction is also the extraordinary ray direction, *i.e.,* å. However, in biaxial minerals, particularly, triclinic minerals, no such specific coincidence is present (Chapter 4, page YY). When the cleavages are seen, then cleavage traces may be determined to be length fast or slow. But, again, in many cases, cleavage traces may make a high angle (say, ~ 45°) with optic directions, then, for such a section, it is meaningless to make a statement about traces being fast or slow. These situations are shown in Fig. 7.9. Fig. 7.9A is length slow uniaxial mineral, *i.e.,* uniaxial positive; with ù and å reversed in number, Fig. 7.9B represents a uniaxial negative mineral, *i.e.*, length slow. Figs. 7.9C and D are biaxial orthorhombic length slow and fast respectively. You must understand, here, that in biaxial minerals the sign of elongation is not related to positive or negative character. In Chapter 4, (page YY) we have already explained that the positive/negative character in biaxial minerals is determined on the basis whether ã is an acute or obtuse bisectrix. When cleavages are slightly inclined as in the case of Fig. 7.9E, then it is possible to ascertain whether cleavage traces are fast or slow, but for high inclination like that in Fig. 7.9F, it is not possible to say anything definite.

In order to ascertain sign of elongation, help is taken of an accessories, a small device that fits in a slot provided by the manufacturer of all polarizing microscopes. This slot is located above the objective of the microscope, or less commonly below the stage.

There are number of accessories available with the vendors. Basically, these are fragments of a birefringent mineral, but very precisely oriented and of a specific thickness. The most common are the following:

1. Quartz wedge
2. Quarter wave mica plate
3. Unit retardation plate
4. Berek Compensator

Quartz Wedge

We all know that a euhedral crystal of quartz is elongated along the *c*-axis. Being a uniaxial mineral, its c-direction coincides with its optic axis, and being optically positive

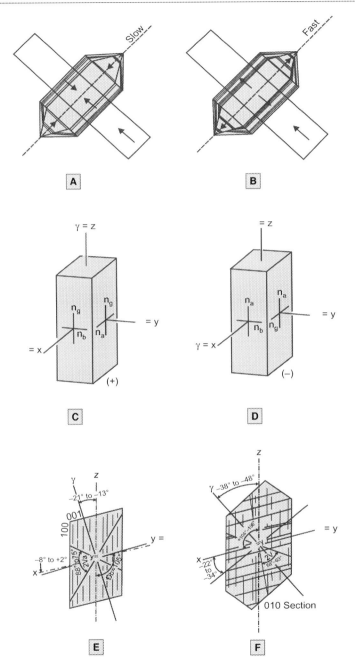

Fig. 7.9 Sign of Elongation. **A.** Uniaxial Positive crystal; **B.** Uniaxial Negative crystal; **C.** Orthorhombic Positive crystal; **D.** Orthorhombic Negative crystal; **E.** Optic direction and cleavages make a small angle making it possible to state sign of elongation; **F.** Optic direction and cleavages make a large angle making it meaningless to state sign of elongation.

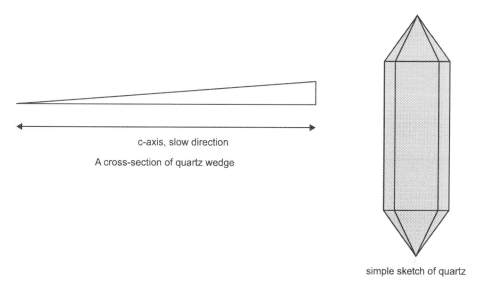

Fig. 7.10 A cross-section of a quartz wedge and a common quartz crystal.

c-direction is slow, *i.e.*, quartz is a length slow mineral. If a slab of quartz is cut parallel to c-axis in such a way that it looks like a wedge as shown in Fig. 7.10; this figure looks exactly similar to the one seen by you earlier (Chapter 4, page YY) where it was used to develop the concept of birefringence and Michael Lévy Chart. As a device it is encased in a metal housing with glass windows along the length of the wedge (Fig. 7.11). The size of the metal housing is such that it fits well into a slot provided above the objective. The slot is oriented at 45° to the field of view. The choice of quartz as the wedge material is dictated by its favourable properties of low birefringence and practically no dispersion of the birefringence. The wedge is inserted in the slot very gradually and always in the crossed polar position; this is important as you will note in the following lines. If you are a beginner then the use of wedge is fun as well as a learning experience. The first step is to make sure that polars are crossed and the field of view is dark. With crosswires properly focused, pay attention to the centre of the field of view. As you gradually push the wedge, the colors of Newton's scale pass through the centre of the view beginning from the first order grey. Depending upon the manufacturer and the price of wedge you would notice up to third, fourth, or higher order colors. A practice of this exercise daily for a few weeks will enable you to recognize various colors of Newton's scale.

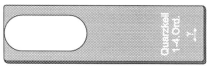

(Courtesy: Leica Co.)

Fig. 7.11 A photograph of a Leica quartz wedge.

Since the quartz wedge is cut parallel to its slow direction we may call it a length slow plate; in other words, the wedge introduces a path difference in the path of light commensurate with this orientation and the corresponding thickness. This path difference will increase if a mineral grain is so placed that the slow direction of the mineral coincides with the slow direction of the wedge. Consequently, the polarization colors shown by the combination of the mineral plus wedge will be of higher order than that shown by the mineral alone. This permits us to mark the slow direction of the mineral; obviously, the direction normal to it will be the fast direction. In order to perform the exercise we need to take the following steps:

1. Select a grain and bring it to the centre of the field of view; a euhedral or subhedral grain should be preferred as its orientation is established easily. If an anhedral grain shows cleavage traces then it may also be selected for the same reason (Fig. 7.12). Avoid basal sections of uniaxial minerals and sections normal to one of the optic axes of biaxial minerals. Optic orientation sketches given in Part II of this book would help you in this regard.

2. Bring the mineral to its extinction position. As you have already learnt, in this position, the vibration directions of the grain are now parallel to those of the polars of the microscope.

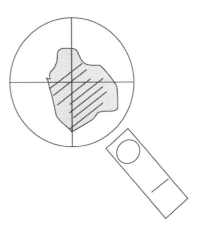

Fig. 7.12 Determination of sign of elongation.

3. Rotate the stage by 45° (with polars remaining crossed, *i.e.*, 45° of the grain); maximum polarization colors are seen in this position; moreover one of the vibration directions of the mineral is now brought parallel to the slot direction, *i.e.*, the wedge direction.

4. Now carefully and slowly introduce the wedge and note change in the order of the polarization color of the mineral. If it rises then the slow of the wedge is coinciding with the slow of the mineral; if there is fall in the order of the color then fast of the mineral is coinciding with the slow of the mineral.

5. Cross check your observation by withdrawing the wedge and rotating the mineral by 90°. Now the second vibration direction of the grain is parallel to the slot direction and the opposite effect to the one observed in step 4 should follow.

Birefringence of a grain can also be determined with the help of the quartz wedge. A principal section of a uniaxial mineral or an optic axial plane of a biaxial mineral is selected for this purpose. If such a grain is not available, then a random section is selected; but in such a case, only the apparent birefringence is determined, which is

Microscopic Examination of Minerals II

always less than the true birefringence. The grain is brought to 45° position and the quartz wedge is inserted slowly. In case the order of colors is falling, then the wedge is pushed till the grain becomes extinct. In such an event, the thin section containing the mineral grain is taken away, and the now the wedge is viewed without disturbing its position. Since the grain had become extinct before we removed it from the stage, the path difference introduced by it is exactly equal and opposite to that introduced by the position of the wedge. The color of that position of the wedge is noted and recorded, and later on compared on the Michael Lévy Chart where numerical value of the birefringence responsible for that color is also noted. It is assumed that the thickness of the grain is 30mì (0.03mm).

Quarter Wave Mica Plate

It is essentially a cleavage fragment of muscovite that is free of any inclusion or fracture. When you will read the description of optical properties of muscovite in the next section (Chapter 14, page YY) you will note that it is a monoclinic biaxial negative mineral with its acute bisetrix (*i.e.*, á-direction) nearly normal to its basal cleavage (*i.e.*, (001) plane). In other words, the cleavage fragment consists of â and ã directions. The thickness of the cleavage fragment of muscovite is so chosen that the path difference introduced by the grain is about 150mì , *i.e.*, ë/4 of yellow light. Because of this retardation it is also called ë/4 plate. The transmitted beam emerges as circularly polarized. The cleavage fragment is encased in a metal housing just like the quartz wedge except that the glass window is much smaller and is circular in shape (Figs. 7.13A and B) with slow direction

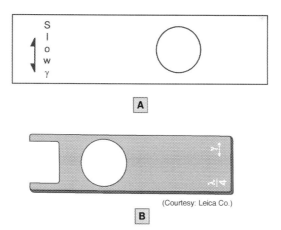

Fig. 7.13 A. A sketch of quarter wave mica plate; **B.** The quarter wave mica plate.

engraved. When the mica plate is view alone under crossed polars a pale grey tint fills the field of view. It is because of this tint it is also called *tinted plate*. The use of this plate is similar to that of quartz wedge, *i.e.*, to determine slow and fast direction of a given grain. Once we reach the step 3 mentioned in the previous section (bringing the grain to its 45° position) the order of polarization colors are noted. The quarter wave mica plate is then introduced; since it is not a wedge the gradual insertion aspect may not be followed in this case. This plate once in place creates a quarter wave path difference over and above that the mineral grain has already introduced. If the fast direction of the mineral coincide with the fast of the plate, the order of the colors will increase; the reverse will take place if the fast direction of the mineral coincides with the slow of the plate. In the Michael Lévy Chart (*see* flap at the end of the book) the movement of the color is easily seen as the color shifts to quarter wave distance high or low depending upon the nature of the mineral. The effect is well noticed if the mineral has high birefringence.

Unit Retardation Plate

As the name suggests, this plate introduces a path difference of one wave length (yellow light) amounting to about 575mì between the two rays emerging out of the plate. It produces a violet tint under the crossed polars; for this reason it is also called a ë tinted plate. Its design is exactly similar to that of the mica plate shown in Figs. 7.14A and B except that the mica grain, if used, is of sufficient thickness so as to create a one lambda

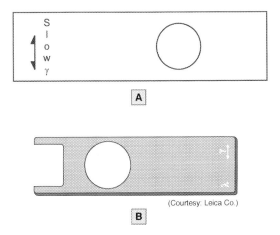

Fig. 7.14 A. A sketch of Lambda Gypsum Plate; **B.** A photograph of Lambda Gypsum Plate.

path difference instead of quarter lambda. Commonly though, either quartz or gypsum is used. Gypsum is easier to use as cleavage fragments of clear transparent gypsum (variety selenite) are easily available; the cleavage fragment tabular on (010) contains á and ã directions so that maximum birefringence is available (Chapter 16, Fig. XX). The application of the unit retardation plate is almost identical to that of quarter plate; the major difference is that the former is more useful for minerals that have low birefringence. For example, quartz shows grey interference color of first order. When examined with unit retardation plate in place, the grain would show an orange of the first order when fast of the mineral coincides with the slow of the plate; blue of the second order is seen when the fast directions of the two coincide.

These plates are also useful in the determination of extinction angle of minerals. When the mineral is perfectly extinct, with the quarter plate in, the field of view will show pearl gray color of the plate. In the case of unit retardation plate in place, the extinction position of the mineral grain will reveal the violet color.

Many workers have noticed that in photomicrographs, many a texture is very conspicuous if photographed with one of the tinted plates in place.

Berek Compensator

Many devices have been proposed for continuously generating compensation in phase difference or retardation introduced by the mineral. The most popular is *Berek Compensator* (Fig. 7.15), which is a simple accessory of the petrological microscope, and is quite useful in measuring birefringence, sign of elongation etc. Its construction is similar to that of a quartz plate; it fits in the same slot. The metal housing has two drums attached at one end with graduation in degrees marked; one of these drums is a vernier drum to the other that yields main division reading.

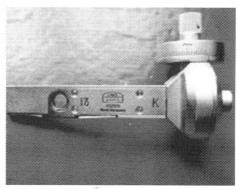

(Courtesy: K.K. Singh and Department of Geology Collection, DU)

Fig. 7.15 Berek compensator.

The drum can be rotated both clockwise and anticlockwise and as the drum rotates a thin calcite crystal attached at the other end of its axle also rotates. At zero reading of the vernier the Berek Compensator shows pearl grey color under crossed polars with no mineral grain placed in the path of the light; this is because the thickness of calcite used in this devise is such as to produce that order of interference color. When we rotate the drum the path of light through the crystal becomes longer, and consequently the retardation changes (*see* Equation XX). Hence, rotation of the drum will bring all Newton's colors in view gradually.

In order to determine birefringence of a mineral, bring the grain to its 45° position, note the color of the mineral. Introduce the Berek Compensator into the slot making sure that it is at its 0° position. Slowly rotate the drum watching carefully whether the order of the colors is rising or falling. If the order is falling, then keep on rotating till the grain becomes extinct. Note the reading on the drum. If, instead of falling, the order of the color rises, then bring the compensator to its 0° position. Rotate the microscope stage by 90°, and then repeat the above exercise to obtain position of complete compensation (extinction position). Now, note the amount of rotation on the drum, this will be in degrees. In the information brochure supplied by the manufacturer of the compensator, there is a table relating this rotation with a numerical value representing retardation. Our advice is that you repeat this experiment by rotating the drum in the opposite direction. In other words, bring the drum reading to 0°, and then rotate the drum anticlockwise, if earlier you have rotated clockwise. You may, alternatively, note the color of the compensator after removing the grain without disturbing the drum reading of the position that caused compensation. Take the help of Michael Lévy Chart to read the value of birefringence of the mineral just as you did it with the help of the quartz wedge. If you are sure that the section has a thickness of 0.03 mm, then substitute in the pertinent formula to directly obtained the birefringence.

In the end the author strongly recommends that each time you undertake a mineralogical or petrological study of thin sections, these accessories should be kept handy for frequent use.

Summary

Following the study of polarization colors, and birefringence etc., the next important property of our interest for observation under crossed polars is the measurement of extinction angle which allows confirmed identification in several cases. From the shape and cleavage traces, and twin planes etc, a guess is made about the crystal orientation of the section. Then this crystal plane is aligned parallel to one of the crosswires and reading on the vernier of the microscope stage is noted. With polars crossed, the grain is rotated till it becomes extinct. The new position is read on the vernier and the difference yields the value of the extinction angle.

Twinning is very well seen under crossed polars. It arises because of planar defects in crystals as a response to changed external conditions. The lattice of the crystal gets distorted in a particular manner for adjusting to the external conditions, thereby, creating two or more indicatrices of different orientation in different parts of the grain. This causes different parts of a given grain to become extinct at different positions of the rotation of the microscope stage. For each mineral the twin pattern is characteristic, therefore, helping observer to identify it.

Disequilibrium during crystallization gives rise to non-uniformity of composition within a grain that is manifested as zoning. Zoning may be identified by sudden or gradual change in colour of the grain under plane polarized light, and /or change in birefringence, which is observed under crossed polar condition. An appreciable time interval between two episodes of crystallization results in overgrowth. An overgrowth can be conspicuous even in plane polarized light, though under crossed polars it is easily spotted.

Contd. ...

Microscopic Examination of Minerals II

> Post-crystallization history of a rock may force individual grains to become internally distorted leading to the distortion of the indicatrix. The result is that the original optical properties are camouflaged; the 'new behaviour' is termed optical anomaly.
>
> In order to study birefringence, order of interference colours, extinction angle, sign of elongation etc. certain accessories are used. The principal accessories in this category are quartz wedge, tinted plates (mica, gypsum and quartz) and a number of phase difference compensators, such as Berek compensator. The fundamental principle of these devices is to compensate the path difference introduced by the birefringent nature of anisotropic minerals. The compensation results in the mineral getting extinct when we use quartz wedge and Berek compensator. Berek compensator provides the numerical value of retardation and birefringence. The tinted plates indicate compensation or enhancement of retardation to the extent marked on the plate.

Important Terms

ë tinted plate 124	*Polysynthetic or lamellar twinning* 115
ë/4 plate 123	*Quarter wave mica plate* 123
Berek compensator 125	*Quartz wedge* 119
Complex twins 115	*Re-entrant angle* 115
Extinction angle 106	*Simple twins* 114
Inclined extinction 106	*Symmetrical extinction* 106
Length fast crystal 119	*Twinned crystals* 113
Length slow crystal 119	*Twins* 114
Optical anomaly 118	*Twin law* 115
Overgrowth 116	*Unit retardation plate* 119
Parallel or straight extinction 106	*Zoning* 116

Questions

1. Why should a mineral become extinct under crossed polars?
2. What is meant of the term 'extinction angle'?
3. Describe the procedure for measuring extinction angle of a mineral.
4. Is it possible for a mineral to have more than one extinction angle?
5. Define the phenomenon of twinning.
6. What is the meaning of 'twin law'?
7. Discuss reasons for twinning to be a characteristic optical property of minerals.

8. How will you recognize a twinned crystal from a normal (untwinned) crystal under microscope?
9. What is the difference between overgrowth and zoning?
10. In how many ways one can recognize zoning? Give examples.
11. What is an optic anomaly?
12. Describe construction of a quartz wedge.
13. What are the favourable points for using quartz as the material for construction of a wedge?
14. How do you determine the order of interference color of a given mineral with the help of a quartz wedge?
15. What is the major difference between mica plate and gypsum plate?
16. For what kind of minerals the use of mica plate is preferred over gypsum plate?
17. What is the principle of Berek Compensator?
18. Describe the construction of Berek Compensator.

Suggestion for Further Reading

Winchell, N.H. and A.N. Winchell, 1948, *Elements of Optical Mineralogy, Part I*, Wiley, 5th edn.

Hallimond, A.F., 1956, *Manual of Polarizing Microscope* (Cooke, Troughton and Simms, Ltd.), 2nd edn.

Bloss, F.D., 1998, *Optical Mineralogy*, Mineralogical Society of America.

8

Microscopic Examination of Minerals III: Conoscopic Condition

Learning Objectives

- Concept of interference figure
- Formation of interference figure under different conditions
- Optic character of minerals
- Dispersion phenomenon and optical properties of minerals

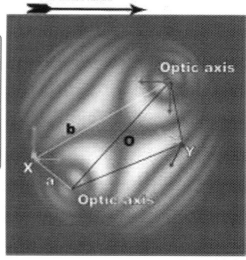

(Courtesy: Leitz Co.)

A photomicrograph of a thin section of a biaxial mineral as seen under conoscopic condition of the microscope. The section has been cut normal to the acute bisectrix and the OAP is at 45° to the polarizer vibration direction.

Prologue

On seashore far a green oak towers,
And to it with a gold chain bound,
A learned cat whiles away the hours,
By walking slowly round and round.
To right he walks, and sings a ditty;
To left he walks, and tells a tale....

— Alexander Pushkin

Chapter's Outline

- The microscope as conoscope
- Uniaxial crystals
- Biaxial crystals
- Dispersion
- Optic sign determination

© The Author(s) 2023
P. K. Verma, *Optical Mineralogy*,
https://doi.org/10.1007/978-3-031-40765-9_8

Introduction

In the previous two chapters, we have used a beam of light that consisted of a bundle of parallel rays that propagated through the microscope tube parallel to its axis. Such an examination, named in the previous chapter, orthoscopic examination of minerals, reveals the nature of a given mineral to a large extent. However, the determination of many optic properties of an anisotropic mineral, *i.e.*, uniaxial or biaxial nature, optic sign and the nature of dispersion, we need to use a conical beam of light. In the present chapter, we shall discuss the concept of convergent light rays that allows us to exactly do such an experiment; we shall relate these observations to the concept of indicatrix that we have already learnt.

The Microscope as a Conoscope

An arrangement is required in the optical system of the microscope in order to generate a cone of light rays, *i.e.*, a conical beam of light produced from a bundle of rays emerging from the illuminator of the microscope. Fig. 8.1 is a sketch of essential elements of a microscope along with a ray diagram depicting the microscope in its conoscopic form. The plane polarized rays from the lower polar (P) are restricted by a *sub-stage iris diaphragm* (I). Between the iris diaphragm and the microscope stage an assembly of lenses (CON) can be brought within the path of rays; this assembly is known as the *condenser* (*see* Chapter 5, page 65). The rays emerging out of the condenser form a conical beam whose apex lies at the object (M) placed on the microscope stage, and this conical beam forms another conical beam, in fact its own mirror image, while traversing through the mineral as shown in Fig. 8.1. The incident cone and the refracted cone share the apex. This conical bundle is made to converge at the *upper focal plane (UFP) of the objective* (O) after traversing the mineral. In this plane every point is the focal point of that ray that has passed in a definite direction through the crystal. The rays enter the upper polar (A) as sets of ordinary and extraordinary rays to, finally, undergo interference of light above the analyzer (A), and an *interference figure* is formed which normally does not lie in the *focal plane of the eye piece*. The image is brought into the focus at C by inserting an additional

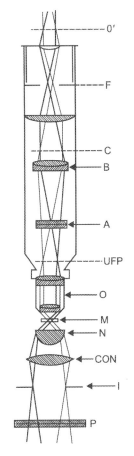

Fig. 8.1 A diagrammatic view of the conoscopic arrangement in a polarizing microscope.

lens in the path of the light above the analyzer. This lens is called the *Bertrand lens* (B). It will be prudent to compare this diagram with that in Fig. 5.1 in Chapter 5, which is the illustration of formation of an object image of the mineral grain. The interference image is different from the object image as you will see in this chapter.

In order to understand the behaviour of light consisting of conical beam of rays through an anisotropic crystal let us illustrate the phenomenon with the help of an example of a uniaxial crystal. The mineral section, ABCD, in Fig. 8.2, is cut normal to the optic axis and receives a beam of light rays in the plane of the paper that also contains the optic axis as shown. We know that any ray propagating parallel to the optic axis will not suffer double

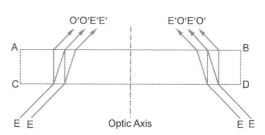

Fig. 8.2 Passage of a conical beam of light rays parallel to the optic axis of a uniaxial crystal.

refraction; however, in the present case we have conoscopic arrangement that produces a series of rays inclined to the optic axis, therefore, each ray of this conical beam, with the exception of the one along the optic axis, will suffer double refraction, *i.e.*, will be split into extraordinary and ordinary rays. Each point on the surface AB will be the emergence of both extraordinary and ordinary rays though one of them would follow the other because of the velocity difference. When the path difference between the two rays is ë or its integral multiple the rays suffer destructive interference (Chapter 1, page YY). This phenomenon is true for each ray of the conical beam; resulting beam is also conical in nature except that at every ë interval, a conical surface will have no light (darkness). When viewed normal to the optic axis the 2-D planar surface will present circles of darkness as in Fig. 8.3A. Along the N-S and E-W directions, which are vibration

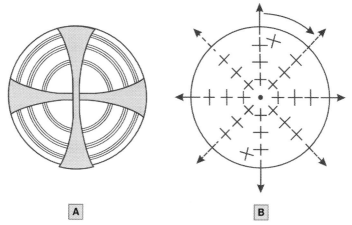

Fig. 8.3 A. Centred uniaxial interference figure; **B.** Schematic representation of extraordinary and ordinary ray directions in the interference figure.

directions of the lower and upper polars respectively, there will be no splitting of rays, and, so these directions are marked by dark bands in the form of a dark cross. These bands are called *isogyres*. If we rotate the nicols, but still maintaining orthogonal relationship between their vibration planes, these isogyres will rotate in the same sense; but rotation of the microscope stage, leaving the vibration directions of the polars unchanged, does not produce any rotation of the figure. If the light source is sunlight (polychromatic beam) then color bands (following Michael Lévy chart) appear. These bands are called *isochromes*. The point of intersection of isogyres is called *melatope*. This figure (Fig. 8.3) is the interference figure that is viewed at C of Fig. 8.1. Fig. 8.3B explains the disposition of vibration directions in the interference figure. From this you will notice that radial directions represent vibrations of the extraordinary rays, and the tangential directions that of the ordinary ray.

Uniaxial Crystals

There are three ways in which a uniaxial mineral can be viewed under the microscope set up for conoscopic observations. One is normal to optic axis (also called basal section), the second is a section cut parallel to optic axis, and finally, a random section, which is inclined to optic axis.

Basal or Centred Optic Figure

This has been described in detail in the previous section, and, explained with the help of Figs. 8.3 and 8.4. The centre of the figure, *i.e.*, intersection of the isogyres will coincide with the centre of the field of view of the microscope. The width of the isogyres will be in inverse proportion to the birefringence, *i.e.*, lower the birefringence, broader though diffused, will be the isogyres for example in quartz (Fig. 8.4A). Likewise, the colored bands around the centre will be broader, therefore, only a few would be seen in the field of view. In contrast, the high birefringent minerals, like calcite, will illustrate a centred

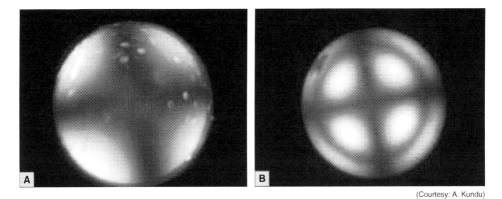

(Courtesy: A. Kundu)

Fig. 8.4 The interference figure. **A.** Quartz; **B.** Calcite.

figure with sharp and thin isogyres and numerous colored bands (Fig. 8.4B) will form. Note that isogyres and isochromes are the result of interference of light, hence, undoubtedly, depend upon the birefringence of the mineral, but also, depend upon thickness of the mineral grain in accordance with the formula given in Chapter 4 (page YY).

Sections Parallel to the Optic Axis (Optic Normal Section)

Such sections are frequently met with in rock thin sections since most tetragonal and hexagonal minerals tend to align themselves along the *c*-axis, which is the optic axis. When such a section is viewed under conoscopic conditions, we do not observe isogyres like the ones observed in the basal section, instead, diffused hyperbolics may be observed. More commonly, we observe colored bands that are restricted to individual quadrant as shown in Fig. 8.5; with the rotation of the microscope stage the colored bands also move. At a given stage of rotation, the hyperbolic isogyres would enter the field of view only to disappear within a few degree of rotation of the stage. For their quick entry and exit from the field of view this interference figure is called the *Flash Figure*.

Fig. 8.5 is oriented in the 45° position. In this position it is easy to decipher the direction of the optic axis that lies across the colored bands that show decreasing order of interference colors counting from the centre of the field of view. The explanation of this phenomenon lies in the fact that as the optic axis is approached the birefringence decreases causing decline in the order of the interference. Flash figures are more prominent when the birefringence is low; in highly birefringent minerals it becomes very difficult to predict nature of the order of the colors.

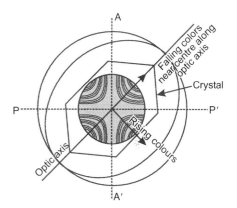

Fig. 8.5 Flash figure in a uniaxial positive crystal.

Random or Oblique Sections

A mineral grain cut neither parallel nor normal to optic axis is called a *random section, oblique section, or off-centered section*. The interference figure obtained from such a section will be off-centered in the field of view; the distance between the intersection of crosswires and that of the isogyres of the figure depends upon the angle of the optic axis of the mineral and the axis of the microscope (Figs. 8.6A-D). Such off-centred figures do not remain stationary when the microscope stage is rotated but make a crankshaft kind of motion. If the angle between the optic axis and the microscope axis is large, then

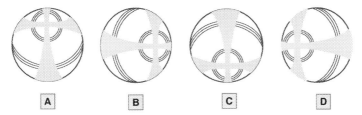

Fig. 8.6 Eccentric Uniaxial Figures. **A.** 0° position; **B.** at 90°; **C.** at 180°; **D.** at 270°.

the intersection of the isogyres lies outside the field of view resulting in only one isogyre being observed (Figs. 8.7A-D).

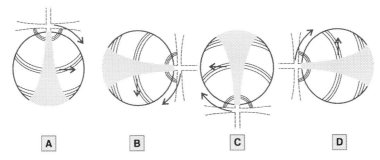

Fig. 8.7 Highly Eccentric Uniaxial Figures; in such cases only one isogyre is seen. **A.** 0° position; **B.** at 90°; **C.** at 180°; **D.** at 270°.

Biaxial Crystals

There are five different orientations of a biaxial crystal with reference to its optic axes. A biaxial mineral grain may be cut *normal to its acute bisectrix, normal to its obtuse bisectrix, normal to one of the optic axes, parallel to its optic axial plane (OAP), or a random section.*

Sections Normal to the Acute Bisectrix

The acute bisectrix (Chapter 4) is that optic direction, abbriviated Bx, which bisects the acute angle between the two optic axes of a biaxial indicatrix. By design of the indicatrix, this optic direction is either ã or á. The optic axial angle that this direction bisects is termed $2V_{ã}$ or $2V_{á}$. With reference to Fig. 4.9 we know that a section cut normal to a bisectrix will contain â direction. Within this plane the direction normal to â will either be á or ã direction depending upon which one of these two is the acute bisectrix. For viewing the interference figure in a biaxial mineral a slight modification in the procedure is made in the sense that the grain is brought to its extinction position first, and then the Bertrand lens is introduced. An interference figure is observed that resembles the one obtained with a uniaxial mineral, but with a notable difference that the isogyre aligned N-S is slightly broader as compared to the E-W one (Fig. 8.8A). The important difference from that of the uniaxial figure is observed when we rotate the microscope stage. The isogyres no longer remain crossed but 'separate out' till the grain acquires 45° position (Fig. 8.8B); further rotation initiates the 'closing in' phase of the isogyres that ends with the 90° position when the isogyres are crossed again. In addition to the isogyres, the figure also contains colored bands if white light is used or dark bands when monochromatic rays are transmitted. But the arrangement of bands is different since it is not concentric; rather two small circles appear on two sides of the E-W isogyre, which is also the thinnest at the centre of the circles (Fig. 8.8). In the 45° position these move to opposite quadrant as shown in Fig. 8.8B. The higher order bands become pear shaped and still higher ones merge with the other bands from the opposite quadrangle. The centres of these circular bands are called *melatopes* and represent the emergence of the two optic axes.

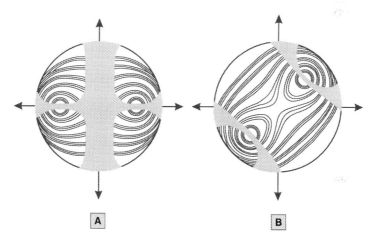

Fig. 8.8 Biaxial interference figure. **A.** A biaxial centred interference figure in the initial position; **B.** The biaxial interference figure in 45° position.

The biaxial interference figure obtained in Fig. 8.8 is easily explained by the same concept that was used to explain uniaxial figure. In the extinction position, the OAP containing the two optic directions is parallel to one of the crosswires. But in the conoscopic arrangement this condition is met only in a narrow regions in the vicinity of the crosswires; rest of the field of view is a region of double refraction. The two rays produced will travel with different velocities and the path difference between the two is controlled by the inclination of the rays to the microscope axis and the two optic axes. Therefore, unlike the uniaxial case, the biaxial minerals will generate two cones one each around each optic axis. The arrangement is shown in Fig. 8.9. In the 45° position we have rotated the OAP away from the vibration directions of the polars, hence, the field of view is illuminated leaving only a part of the conical arc associated with the optic axes.

The maximum separation of the isogyres is attained when the stage is in 45° position. The actual distance between the two melatopes is a measure of the optic axial angle (2V) for a given â, and for a given objective – ocular system of a microscope. The number of wavelength retardation is equal to the number of dark bands you see between either of the melatopes and the centre of the field of view. Because of the refraction in air, the emergence points of the optic axes, i.e., the melatopes may not appear situated at their true location. In other words, we normally end up measuring, what is called the *apparent optic axial angle, 2E* (Fig.8.10). The angles E and V are related to each other by the following relationship:

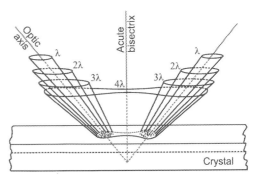

Fig. 8.9 Explanation of biaxial interference figure; cones represent surfaces of equal retardation.

$$\text{Sin } E = \text{â sin } V$$

The NA (numerical aperture) of the objective determines the maximum admissible cone of rays for which both melatopes will lie in the field of view. As an example we may take a case of â = 1.5, then a 2V of 70° will mean that the two isotopes lie just at edge of the field of view. From the sine relationship it is obvious that progressively smaller angles are accommodated as â increases. Use of an oil immersion objective (Chapter 5, page 65) allows the light to enter the objective without refraction; hence, true 2V is nearly seen. This also means that the use of an oil immersion objective makes it possible to see both melatopes for comparatively large 2V.

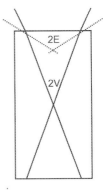

Fig. 8.10 Real and Apparent Optic Axial Angle.

Sections normal to the Obtuse Bisectrix

When a thin section is cut normal to an obtuse bisectrix it means that the two optic axes are wide apart and their emergence may not lie within the field of view. However, when such a grain is brought to its extinction position, and then viewed under conoscopic condition the interference figure is exactly similar as mentioned above, *i.e.*, two isogyres forming a cross like the one shown in Fig. 8.8. However, the isogyres rapidly move away from each other as the stage is rotated by a few degrees only; the rotation, in some cases is so much rapid that much before the 45° degree position, the melatopes are out of the field of view. Because of the melatopes being out of the field of view, it is not possible to get a reasonable idea of the 2V, which was possible in the case of an acute bisectrix normal section of the previous section. But other features are common such as the arrangement of color bands, â direction normal to the line joining the melatopes etc. (Fig. 8.8).

Sections Parallel to the OAP

A section that is parallel to the OAP of a biaxial mineral is also normal to the â direction; such a section contains both á and ã directions. Optic axes also lie in this plane causing the cones of successive path difference to be four sets of hyperbolic traces as shown in Fig. 8.11. The two opposite quadrants carry one of the bisectrices. The arrangements of color bands is such that the quadrants associated with the acute bisectrix exhibit lower order colors at a given distance from the centre of the field of view as compared to the ones that carry obtuse bisectrix. This distinction, between the orders of interference color of the two pairs of opposite quadrant, increases as the value of 2V increases from 90°.

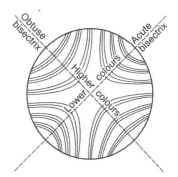

Fig. 8.11 Optic figure in a section cut parallel to OAP of the mineral grain.

Sections Normal to an Optic Axis

When a grain of a biaxial mineral is cut normal to one of optic axes, the section will exhibit the emergence of this optic axis at the centre of the field of view. There will be one dark isogyre (Fig. 8.12) when the grain is viewed under the extinction position; the isogyre will be without curvature as shown in Fig. 8.12A but will become curved in the 45° position (Fig. 8.12B). The curvature of the isogyre in the 45° position is inversely proportional to the value of 2V. When 2V has a value of 90° the isogyre will remain

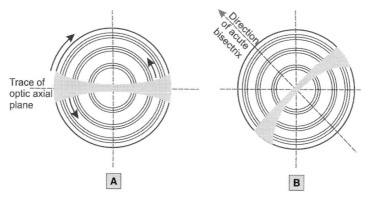

Fig. 8.12 A. Optic figure in a section cut normal to one of the optic axes of the mineral when the mineral is in extinction position; B. The above figure now after the mineral is rotated to 45° position.

straight; on the other hand, if the 2V is very small the second optic axis may emerge near the periphery of the field of view.

Such sections that are normal to one of the optic axes are easily recognized under crossed nicols because the grain in such a situation will be uniformly illuminated upon rotation of the stage.

Random Sections

When a grain is cut oblique to OAP, the section so obtained is termed a random section, and often this is common when the habit of the given mineral is tabular on a face which is not its OAP. If the 2V is sufficiently large then the interference figure would shift the location of one of melatopes. Now, this melatope will lie off-centered from the centre of the field of view. In order to understand this let us take an example of an orthorhombic mineral which has its OAP parallel to (010) plane. A prism face, (110) will be an oblique plane from our point of view. A prismatic section, then, will show a lateral displacement of the interference figure as shown in Figs. 8.13A and B. In a second example of the same mineral, let us suppose the section is cut now parallel to (101). In this event,

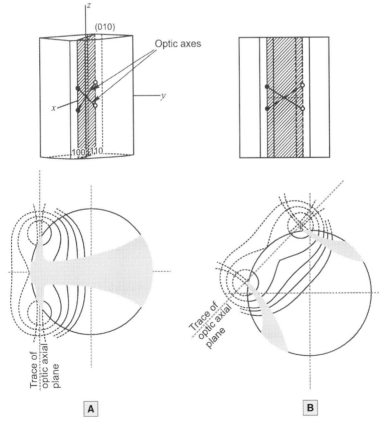

Fig. 8.13 A. A prismatic section (110) of an orthorhombic mineral and the optic figure as seen in extinction position; **B.** The above figure now after the mineral is rotated to 45° position.

we may be able to view only one isogyre for most part of the rotation of the stage (Figs. 8.14A and B). Still a third case, not uncommon in everyday experience, is when the above crystal is sectioned parallel to (111). This may allow only one isogyre to be

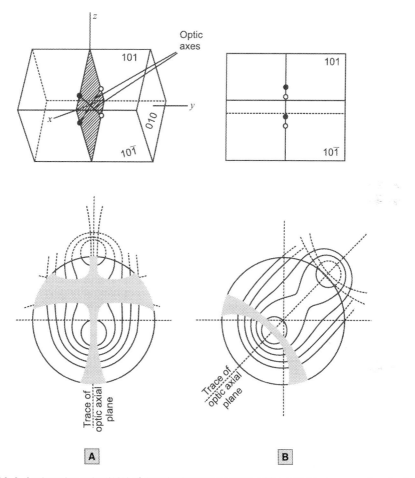

Fig. 8.14 A. A prismatic section (101) of an orthorhombic mineral and the optic figure as seen in extinction position; **B.** The above figure now after the mineral is rotated to 45° position.

viewed at any give instant of rotation of the stage.

Instances like this, when only one isogyre is seen, are not easy to interpret if the mineral is unknown since we may be viewing an off-centered uniaxial or biaxial figure, or a biaxial grain cut normal to one of the optic axes. It is best advised to locate another grain; for this a scrutiny of the grains in the thin section must be made in plane polarized light. The shape of the grain, pattern of cleavages, partings etc. would provide a clue for selecting a right grain-section. However, if we know that the grain being observed is a biaxial one, then the single isogyre may be brought parallel to one of the crosswires. If the isogyre is occupying the centre of the field of view, then it is normal to one of the optic axes, otherwise, it is an oblique section as described above.

Dispersion

In optics, *dispersion* is the phenomenon in which the phase velocity of a wave depends on its frequency. This phenomenon has been explained to you in school curriculum with the help of a prism that splits the sunlight into a spectrum of seven colors, popularly called VIBGYOR. You also learnt that refractive indices decrease with increasing wavelengths. It is termed *normal dispersion*. For a clear understanding of this concept, we must make a reference to the nature of light propagation. In Chapter 1 we mentioned, while discussing nature of light, that electron cloud associated with each constituent atom in a crystal has characteristic natural resonant frequencies; one of them is in the light frequency range though both may not exactly coincide. The impact of an incident light photon sets off the light frequency resonance in the electron cloud causing the atom to become a source of light with a phase difference. This difference in phase is a function of the degree of difference in frequencies between the incident photon and the natural frequency of the cloud. When the frequency difference is sufficiently large dispersion would occur generating normal dispersion. On the other hand, if both frequencies are same then strong absorption will result, obviously, decreasing the refractive index of the medium. Such a medium shows increase in index with increasing wavelength. This is known as *abnormal dispersion*.

When refractive index of a mineral is reported in a textbook, such as the present one, it is customary to use the value for the sodium light *i.e.*, yellow component of the visible spectrum. One obvious reason is that it lies nearly in the middle of the visible spectrum, but, equally important is that its wavelength, 589nm coincides with one of Fraunhofer lines (Technical Text Box 3.1).

From the optical mineralogy perspective the importance of dispersion lies in the fact that we use white light to study optical properties of minerals. This beam consists of seven prominent wavelengths. For each wavelength associated with a beam of light rays, there are that many indicatrices because an indicatrix is defined by the length of its semi-axes that are proportional to the refractive indices. In other words, the grain will exhibit dispersion when the interference figure is obtained from an anisotropic mineral. We all know that in the case of a uniaxial mineral there is only one optic direction, *i.e.*, the c-axis or the z-direction along which there is no double refraction (Chapter 4, page 50). This character of uniaxial mineral is fixed and does not change with chemical composition or the wavelength of the transmitted light ray. Hence, the dispersion is not effective in the uniaxial minerals because each wavelength produces an indicatrix that shares the same optic axis direction. But the situation is different for biaxial minerals where the dispersion can be conspicuous and may create confusion if not properly accounted for. The refractive indices, á, â, and ã will be different for each wavelength (marked by a particular color) as we learnt from Chapter 1 (page 6); moreover, the values do not necessarily vary systematically and proportionately for each wavelength. Physically noticeable effect of this variation is manifested in the variation in the value of optic axial angle, 2V. In effect, one may say that biaxial minerals exhibit a *dispersion of optic axes*. Depending upon the value of 2V of a given mineral, a given color may show obtuse 2V and another color of the same beam may show acute 2V. Furthermore, in monoclinic and triclinic crystals, unlike uniaxial and orthorhombic minerals, the orientation of the bisectrices relative to the crystal axes is different for each wavelength causing a *dispersion of the bisectrices*. In the context of its importance and implications, let us discuss dispersion by biaxial minerals belonging to each crystal system.

Orthorhombic Minerals

You have learnt that for an orthorhombic crystal á, â, are ã are coincident with *a*, *b*, and *c* (not necessarily in this order) (Chapter 4, page YY), which means that the optic bisectrices, á and ã, are parallel to two of the three crystallographic axes. This will be the case for all the colors. A hypothetical case is shown in Fig. 8.15A where OAP is parallel to (100) plane and *c*-axis is the acute bisectrix. Let us, further assume that the 2V subtended by red wavelength is greater than that by blue (or violet) as is shown in the same figure. To simplify our treatment we take that only these two colors are present in the beam of light. Then, it follows that there would be two isogyres; one for which red is extinguished, and the second for which blue is extinguished. The result would be as shown in Fig. 8.15B where a dark band is fringed outside by the red (because blue is extinct) and is fringed inside by the blue (Fig. 8.15C). The color bands around the melatope will also follow the same pattern if the optic axis dispersion is distinct. The dispersion is stated as *r* > *v* for this example. In the reverse case where 2V of blue (or

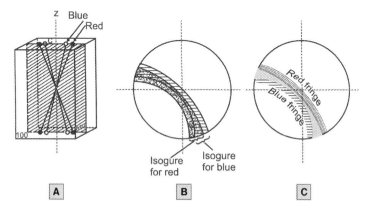

Fig. 8.15 A. An idealized orthorhombic crystal with OAP and two set of optic axes for two different colors; **B.** Isogyres in the normal to acute bisectrix section in 45° position for the above crystal; **C.** Colored fringes indicating *r* > *v* as seen by red fringes on the convex side of the isogyre and violet fringe on the concave side. Reverse would be the case for *v* > *r*.

violet) is greater than that for the red, we shall witness just the opposite phenomenon; and, then the statement would be $v > r$.

Hence, in orthorhombic crystals there is no dispersion of the bisectrices, but only of the optic axes.

Monoclinic Minerals

Monoclinic crystals are a bit complicated optically (Chapter 4, page 60), since only one of the three optic directions is coincident with the *b*-axis which is normal to the (010) symmetry plane for most common monoclinic minerals. A specific direction, á, â, or ã will occupy y, (*b*) direction for individual monoclinic minerals. We also know (Chapter 4, page 60) that OAP is normal to â; hence, with regard to crystal orientation three cases arise. The first is when â is coincident with b axis, second is when b axis is an acute bisectrix; finally, the last case when b is an obtuse bisectrix. The three cases are sketched in Fig. 8.16.

In the first case when â coincides with *b*-axis, the OAP for all wavelengths will be

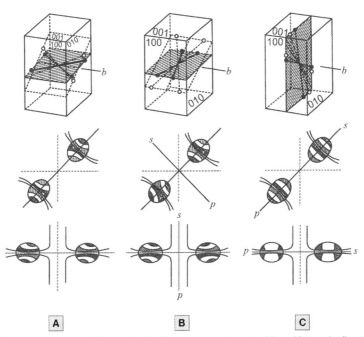

Fig. 8.16 Dispersion by monoclinic crystal. **A.** Case I. y direction coincides with â optic direction so the OAP becomes (010) for all wave lengths; **B.** Case II. y direction coincides with the acute bisectrix; **C.** Case III. y direction coincides with the obtuse bisectrix.

Microscopic Examination of Minerals III: Conoscopic Condition

within the (010) plane. This means that both optic axes and the bisectrices lie within (010), and will be dispersed parallel to this plane. A conspicuous manifestation of such dispersion is that the color bands around one melatope is elongated as compared to the other one. For this reason, it is called *inclined dispersion*.

In the second case where b axis is an acute bisectrix, the obtuse bisectrix and â lie within (010) plane. The dispersion will affect the OAPs of different wavelengths though the bisectrix (á or ã) that coincides with b axis would continue to be normal to OAPs. The melatopes rotate around this bisectrix giving rise to a *centrosymmetrical interference figure*. The dispersion is termed *crossed or rotated dispersion*.

In the third case, it is the acute bisectrix that lies in the (010) plane along with â for all colors. The acute bisectrix is displaced towards the line that is normal to the line joining the melatopes as shown in Fig. 8.8. Because of the dispersion of the optic axial angle, a succession of melatopes occur parallel to each other; for this reason this dispersion is called *horizontal or parallel dispersion*. We can easily distinguish such dispersion in the isogyre crossed position as the line joining the melatopes is laced with different colors on both sides.

Triclinic Minerals

We know that optic orientation of triclinic minerals involve non coincidence of crystal axes, a, b, and c, with optic directions, á, â, and ã. The dispersion, therefore, is irregular, and no prediction about the behaviour of color bands of an interference figure can be made.

Crossed Axial Plane Dispersion

You have been told earlier that â is intermediate between á and ã in its values, but not necessarily the arithmetic average of the two extreme refractive indices. In fact, as a rule â is closer to either á or ã. When â is appreciably close to one of the extreme values, the variation in 2V for each color is rather very conspicuous. For a normal to acute bisectrix figure, 2V rapidly changes with the change in the wavelength; as the wavelength decreases, 2V decreases, and for a given wavelength it may become zero, *i.e.*, the grain behaves like a uniaxial mineral. Further decrease in wavelength results in opening of the optic axis but with OAP now oriented normal to that of the previous situation. This is called the *crossed axial plane dispersion*. In such cases, (*e.g.*, brookite) the interference figure does not carry perfectly dark isogyres.

Optic Sign Determination

The determination of positive or negative character of an anisometric mineral is useful not only for the sake of identification, but, also, to get an idea of its chemical composition if the mineral represents a binary solid solution. For example, forsterite is biaxial positive but fayalite is negative. When the grain has acquired euhedral outline, then the use of compensators would indicate about its positive or negative character, particularly in the case of a uniaxial mineral (Chapter 7, page YY). However, for anhedral grains or grains that don't present undoubted outlines, interference figure is the best method to determine its positive and negative character.

Uniaxial Minerals

The first step towards sign determination of a uniaxial mineral is to select a basal section of the mineral from amongst those available. A basal section of a uniaxial mineral presents extinction throughout the rotation of the stage; a euhedral grain can be recognized by its outline. The conoscopic examination reveals a centred figure as we already know from the previous pages of this chapter. We also know that radial directions represent the direction of vibration of the extraordinary rays and the tangential directions are those of the ordinary ray. In the next step we insert an accessory plate; from the birefringence seen during the orthoscopic examination we already have an idea about the mineral being a high birefringent or low one. If the latter is the case we use unit retardation plate, *i.e.*, gypsum or quartz plate; otherwise we use 1/4 ë plate, *i.e.*, mica plate. The fast direction is known from the engraving on the housing of the plate. So, in the final step we insert the plate and note whether fast of the plate is raising the order of the color or lowering it. A rise in the order of the color bands mean that the radial direction is fast so the mineral is uniaxial negative. Normally, it will mean that the quadrant of the interference figure along which the accessory plate has been inserted, the color closest to the centre will be yellow and in the opposing quadrant it would be blue (Fig. 8.17). It may cautioned here that Fig. 8.17 is true for the arrangement shown, but, if your microscope slot is positioned across the one shown, then, reverse results would be obtained. When ë/4 plate is used total compensation in the form of dark bands may also appear close to the centre. If the centred figure is not obtainable from the given set of grains, then off-centred figure may be used with caution. Beginners should avoid a single isogyre figure. We may use quartz wedge instead of a unit or ë/4 plate, particularly if the birefringence is high. In the case of quartz wedge you have to watch out in which direction the colors are moving in order to decide whether radial direction is fast or slow.

Microscopic Examination of Minerals III: Conoscopic Condition 145

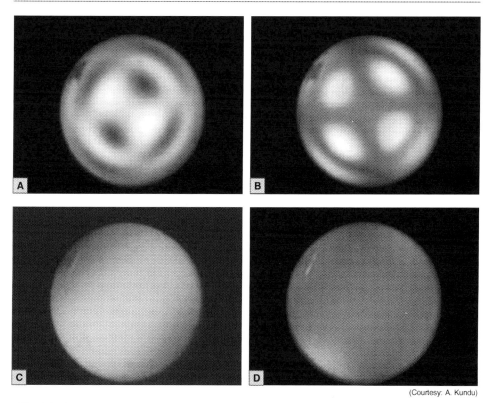

(Courtesy: A. Kundu)

Fig. 8.17 A. Uniaxial positive mineral as seen through ë plate; **B.** Uniaxial negative mineral as seen through ë plate; **C.** Uniaxial positive mineral as seen through ë/4 plate; **D.** Uniaxial negative mineral as seen through ë/4 plate.

Biaxial Minerals

The first step in the process of optic sign determination for a biaxial mineral involves selection of a grain that has been sectioned normal to the acute bisectrix. The conoscopic examination will yield a centred biaxial figure that should be brought to the 45° position at which there is a maximum separation of the isogyres and the melatopes are clearly visible. The line joining the melatopes is the trace of OAP normal to which is the â direction. If the other direction that is along the melatopes is á, then the grain being examined is positive, alternatively, if this direction is ã, then the mineral is negative.

The determination of the nature of melatopes direction is done with the help of the accessory plates just as was the case for the uniaxial mineral. Because the direction normal to the melatopes line is always â, the slow direction normal to it is ã; if it is fast then á is the right answer. For this we insert the ë plate (quartz or gypsum) and note

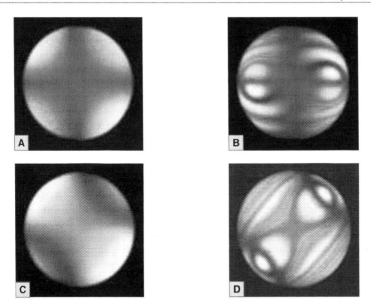

Fig. 8.18 A. Biaxial positive mineral (barite) as seen through ë plate, normal position; **B.** Biaxial negative mineral (muscovite) as seen through ë plate, normal position; **C.** Biaxial positive mineral (barite) as seen through ë plate, 45° position; **D.** Biaxial negative mineral (muscovite) as seen through ë plate, 45° position.

whether the inner side of the isogyres in the vicinity of melatopes is yellow; if yes, then the grain being examined is biaxial positive (Fig. 8.18). The grain showing higher order blue color in the inner side of the isogyres close to the melatopes when ë plate is inserted in biaxial negative mineral.

Summary

In order to study optical properties of minerals under conoscopic conditions, the optical system of the polarizing microscope is slightly rearranged. The condenser is brought in the path of light after the beam is converted to plane polarized. A moderately high power objective is used. This arrangement produces first a converging conical beam that becomes a diverging conical beam within the mineral. An image, called interference image, is formed above the analyzer but is out of focus for the ocular. It is brought in focus by means of Bertrand lens. The formation of the interference figure is explained by means of interference phenomenon between extraordinary and ordinary rays generated by the anisotropic minerals. The dark bands in the figure are

Contd. ...

called isogyres and the colored bands are termed as isochromes. The emergence spot of optic axis is known as melatope. The uniaxial centred interference consists of a dark cross with a number of circular colored bands. Sharpness and width of isogyres and isochromes are related to the birefringence of minerals for a given thickness of the grain. If a uniaxial crystal is sectioned parallel to its optic axis, and then viewed under conoscopic conditions, a series of hyperbolas in each of the four quadrant of field of view are observed. Since these appear to enter and leave field of view in quick succession, the figure is termed as the flash figure. A random section of a uniaxial mineral may make a small departure from being normal, and then it exhibits a displaced melatope, which rotates as the stage is rotated but maintains crossed nature of isogyres. However, when the section makes a large departure from being normal only one isogyre is seen. There are five different possibilities of orientation of a thin section of a biaxial mineral with respect to its optic directions. In order to see a well formed biaxial figure we normally study a section normal to acute bisectrix. In the extinction position a cross consisting of two isogyres is seen but the E-W isogyre is thinner than the N-S one. Moreover, as the stage is rotated the isogyres separate out along with the color bands that form concentric circles, now around two melatopes. The maximum separation is at 45° position, and then these close in to form a cross again. A section cut normal to obtuse bisectrix would show isogyres rapidly moving apart, and melatopes may, actually lie outside the field of view at 45° position. The third type of section is parallel to optic axial plane. In this case the optic figure consists of four sets of hyperbolas that are related to the two bisectrices. A thin section of a biaxial mineral may also be cut normal to one of optic axes; in such cases only one isogyre appears. Finally, a random section may an off-centred biaxial figure with two isogyres that may cross outside field of view; in extreme case only one isogyre may be noticed.

The interference figures obtained above often reveal dispersion of white light into its component. The uniaxial minerals, due to their property of optic axis being parallel to c-axis show minimum dispersion, whereas the biaxial minerals would show several types of dispersion. In orthorhombic minerals the dispersion is of optic axis only but in monoclinic crystals optic axis and bisectrix dispersion are noticed. Dispersion is complicated in triclinic grains.

Interference figures are used in determination of optic sign of uniaxial and biaxial minerals with the help of compensators.

Important Terms

Abnormal dispersion 140

Bertrand lens 131

Centrosymmetrical interference figure 143

Condenser 130

Interference figure 130

Melatopes 135

Normal dispersion 140

Normal to an optic axis 134

Crossed or rotated dispersion 143

Dispersion 140

Dispersion of optic axes 141

Dispersion of the bisectrices 141

Flash figure 133

Focal plane (F) of the eye piece 130

Inclined dispersion 143

Isochrones 132

Isogyres 132

Normal to acute bisectrix section 134

Normal to obtuse bisectrix section 134

Oblique section 133

Off-centered section 133

Parallel to optic axial plane 134

(OAP) section 134

Random section 133

Sub-stage iris diaphragm 130

Upper focal plane (UFP) of the objective 130

Questions

1. What is a conical light beam?
2. How a conical light beam is produced?
3. In what manner the conical beam travel within the mineral?
4. What is an interference image?
5. How does the formation of an interference image differ from that of an object image?
6. Describe essential features of a centred uniaxial interference figure.
7. What is the relationship of birefringence of a uniaxial mineral and pattern of its interference figure?
8. Describe essential features of a biaxial interference figure.
9. How does a biaxial interference figure arise?
10. Describe briefly the types of sections in which a biaxial interference can be studied.
11. How does a medium cause dispersion of white light?
12. What is the difference between dispersion by a biaxial mineral and that by a uniaxial mineral?
13. What is a crossed axial plane dispersion?
14. Give a summary of the method for determining optic sign of a uniaxial mineral.
15. Give a summary of the method for determining optic sign of a biaxial mineral.

Suggestion for Further Reading

Wahlstrom, E.E., 1970, Optical Crystallography, 5th ed., Wiley, New York, 288p.

Hartshorne, N.H. and A. Stuart, 1973, Practical Optical Crystallography, Elsevier, 326p.

9
Reorienting Techniques

Learning Objectives

- Development of methods to achieve desired orientation of mineral grains
- Designing accessory stages and to appreciate their limitations
- Determination of the optic properties of minerals by reorienting techniques.
- Correlation of physical and chemical characteristics with optical properties

Prologue

........From these examples alone, it seems clear that the way to begin an episode of a major discovery in science, at least an observational science like earth science, is through an exploration of a new and important frontier. Once the comprehensive base of fresh new observations is available, a kind of interchange between the inductive and deductive modes can be very beneficial ...

— Jack Oliver

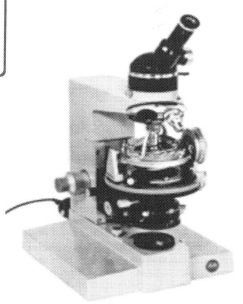

(Courtesy: Leitz Co.)
The Universal stage mounted on a microscope.

Chapter's Outline

- Spindle stage
- Universal stage
- Fundamental principle

Introduction

The early part of the twentieth century was a time of growth of optical mineralogy when scientists fiddled with small crystals in transparent thick liquids in order to bring a desired orientation along the axis of the microscope. Obviously, these techniques were cumbersome and messy; but main hurdle was that the amount of rotation required could not be ascertained or maintained. It was important to develop ability to orient the crystals under microscope, since anisotropic crystals have two limiting refractive indices, and their measurements necessitated two measurements in two different directions. Moreover, with the general acceptability of Fletcher's indicatrix, the biaxial minerals required three measurements in definite directions (Chapter 4, page YY). This led to the design of rotation apparatus in due course. Two such devices are popular and available commercially. One is single axis *spindle stage* that uses a grain mounted at the end of a spindle, and, the other is multiple axes stage which can take up a normal thin section. The latter was developed by *Fedrov* as a *4-axes stage* (General Text Box 9.1), so it is named after him; more popular is the 5-axes stage termed *Universal Stage*. Gradually, other optical parameters, such as 2V, birefringence, extinction angle, optic orientation etc. were also determined by these reorienting equipments fairly successfully.

General Text Box 9.1

EVGRAF STEPANOVICH FEDOROV

There are three variants of the name of this mathematic legend. In Russian, which is his native language, he is spelled as Евграф Степанович Фёдоров, the English spellings are: a) Yevgraf Stepanovich Fyodorov, or sometimes spelled Evgraf Stepanovich Fedorov. E.S. Fedrov was born in the city of Orenburg on December 22, 1853. His was a highly educated family of that time; a number of famous engineers belonged to the Fedrov family that made Saint Petersburg their main residence.

Fedrov's father was an army engineer. He also joined army as a combat officer, but became disinterested that led him to join revolutionary underground movements in late nineteenth century. During this period he travelled extensively and became fascinated by crystallography. Finally, he landed in 1880 to the Mining Institute called Gorny Institute. By 1895, his fame spread far and wide, and he came to occupy distinguished Chair of Professor in Geology in Moscow Agricultural Institute (now Timiryazev Academy).

Contd. ...

Scientists became aware of his remarkable grasp on group theory although he had hardly began his formal scientific education, and was barely 26 years of age. He published his first book, a classic, in 1885 in the form of the 21st volume of Transactions of the Mineralogical Society through encouragement of A.V. Gadolin who derived the 32 classes of symmetry. Fedrov's book systematically proved the existence of 230 space groups. In order to understand the significance of this work, which influenced, Groth, von Laue, Ewald, Bragg, Schoenflies, and countless 20th century scientists, we must take into account the fact that the x-rays were not even discovered by this time! He established the fundamental geometrical rules of the structure of crystals though the crystals themselves remained to be analyzed. Since the only method of such analyses at that time was the optical method and the geometrical analysis with the aid of a goniometer, Fedorov became and remained to the very end of his life the most prominent specialist in the field of petrographic (optical) analysis of minerals. At an amazing speed he brought forth inventions which in themselves would have made his name immortal, *i.e.*, the two-circle goniometer and the Fedorov universal stage. Fedrov and the students close to him collected enormous quantities of data from the measurements of all possible crystals; this material was systematized and published only a year after the death of the author as the monumental volume "Das KristuZriech". From the geometry of crystal, Fedorov derived conclusions about their internal structure, an idea taken up by Harker, Donnay, and others 40 years later. A large group of leading scientists gathered around him from all over the world, *e.g.*, Barker, Braggs, Duparc, Groth, Laue, Ewald etc.

Prof. Evgraf Stepanovich Fedorov breathed his last on May 21, 1919 in St. Petersburg (at the time of his death, it was named Leningrad, but now reverted to St. Petersburg).

The above description is based on an article by I. I. Shafranovskii and N.V. Belov published in *50 Years of X-ray Diffraction* by P.P. Ewald published for the International Union of Crystallography in 1962 by N.V.A. Oosthoek, Utrecht, The Netherlands. Permission to use the material is gratefully acknowledged.

Spindle Stage

Initially, spindle stages were designed in the workshops of various geology departments. Soon, a number of spindle stages became available commercially though many workers prefer to construct their own. Determination of refractive indices of anisotropic minerals and estimation of dispersion by them is best studied by the use of this simple device.

Design of the Equipment

The one shown in Fig. 9.1 is known as the *Wilcox's Spindle Stage*[1]. It consists of a base (metal or plastic) of such a dimension that it could be conveniently placed on the

---

1. Bloss, F.D. 1981. *The Spindle Stage: Principles and Practice*; Cambridge University Press: Cambridge, England.

microscope stage without any corner jutting out. At one end of the base a vertical semicircular graduated plate is glued; transparent plastic is a common material used. A hole near the centre of the semicircular disk carries the spindle made out of a metal wire or a long steel needle bent in L shaped (H in Fig 9.1). Short end of the L is kept outside for the observer to rotate the spindle by an angle that is read on the graduated scale. A thin groove is carved out in the base of the stage so that the spindle fits into the groove; this allows one to control the position of the spindle. To strengthen the spindle position two clamps, like the one used on the microscope stage for holding thin sections, are also provided (C_1 and C_2 in Fig. 9.1). The other end of the spindle is tapered though blunted at the extreme so that a crystal can be mounted with the help of an adhesive.

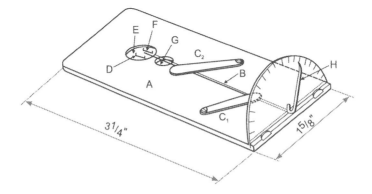

Fig. 9.1 Wilcox's Spindle Stage.

A small glass disc (D in Fig. 9.1) is provided at this end of the spindle in such a manner that the spindle along with the attached crystal gets immersed completely when the disc is filled with a liquid.

Spindle Stage Techniques

The spindle stage has great practical advantage for refractive indices determination of an anisotropic mineral since all the principal vibration directions can be brought in the desired orientation. This is comparatively easy if the crystal is euhedral because a known crystallographic direction is easily spotted and used as the rotation axis. The oriented mounting is of course easier said than done because a slight deviation from the correct orientation may impart a large error in the result. Because of the small size of samples normally the mounting is done under a binocular microscope; even supplementary devices to hold the sample and the needle may have to be fabricated. Once the oriented sample is mounted, and the spindle stage is clamped on to the microscope stage it is ready for the view, under orthoscopic as well as conoscopic conditions. For the refractive index determination the first step is to determine the vibration directions of the crystal;

Reorienting Techniques 153

in other words, ù and å of a uniaxial mineral, and á, â, and ã directions of a biaxial mineral. are located with reference to faces and edges of the euhedral grain. If it is a known mineral, our task is easy since the literature would reveal these optic directions. Most uniaxial minerals are elongated along c-axis, *i.e.*, optic axis (Chapter 4, page YY), therefore, ù and å locations are easy to spot as one of these would be parallel to c-axis. A complete unknown mineral would present problems, particularly, if devoid of any straight edges, cleavages, partings, characteristic striations etc.

Once an optic direction becomes known, then the grain is mounted on the stage with that direction parallel to the axis of the spindle stage. The stage is mounted on the microscope stage and fastened by means of screws provided for this purpose. By making an intelligent guess for its refractive index of the mineral along that particular direction fill the disc with a particular liquid whose refractive index is close to the guessed value. Then follow the steps already outlined in Chapter 3 (page YY) for obtaining Becke line and determination of refractive index. After attaining the Becke line vanishing step, transfer a drop of the liquid to the refractometer. Follow the procedure already given for removing the parallax between the two images in the field of view of the refractometer, and, then, read the value. Repeat the procedure after bringing the second vibration direction along the spindle axis; finally, the third optic direction as well if the mineral is biaxial. These values should be considered as apparent values; repeat this procedure for a slight variation in the direction from the initially selected one at least three times. Select the maximum values for each direction and designate them as ù or å for uniaxial minerals, or á, â, and ã for biaxial minerals.

The method outlined above is simple but hardly practical in most circumstances. For several decades Bloss and students have been improving spindle stage techniques for measurements of optical parameters[2]. Through their efforts, and also because of inherent simplicity of the design of a spindle stage, the use of spindle stage has spread, and, in some respect, overtaken the popularity of U-stage. In particular, small grain size fragments are quite suitable for spindle stage and yield fairly accurate data in comparison to U-stage. Some of the currently popular spindle stages are shown in Fig. 9.2. The spindle stage, in the Bloss method, is used to yield two measurements. One is termed Ms and other S. Ms is the angle read on the microscope stage vernier and denotes extinction angle for a particular value of S, which is the angle read on spindle stage. Commonly, S values begin from 0° and then after 10° increment of S, Ms values are obtained. This allows us to make a table of two columns; one column is for S and another for Ms. The value of angle E is calculated for each value of S by subtracting Ms from 180°. E is that angle which a vibration direction makes with the spindle axis, and this forms the third column in our table. Values are graphically plotted on a Wulf net whose orientation should coincide with that of the spindle stage as shown in Fig. 9.3. Both and E and S values also form as input data set for a computer program called EXCALIBRW, which can be obtained from Evans[3]. The typical output of this program

2. Bloss, F.D. 1999. *Optical Crystallography*; Mineralogical Society of America.
3. clanevans@msn.com

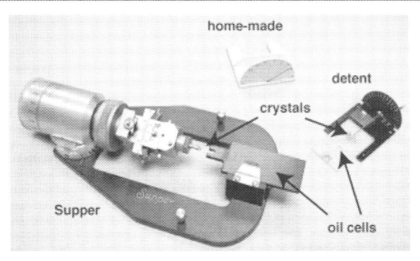

Fig. 9.2 Photograph of three different types of spindle stages and their associated oil cells. The Supper model uses an X-ray goniometer head/brass pin combination to hold the crystal in place. The detent and home-made models can use a needle of some type (*e.g.*, tungsten needle, sewing needle or straight pin) to mount the crystal. (After Gunter et al. 2004)

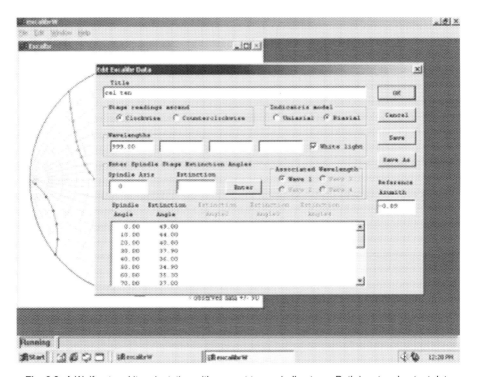

Fig. 9.3 A Wulf net and its orientation with respect to a spindle stage. Both input and output data may be seen. (After Gunter et al. 2004)

Reorienting Techniques

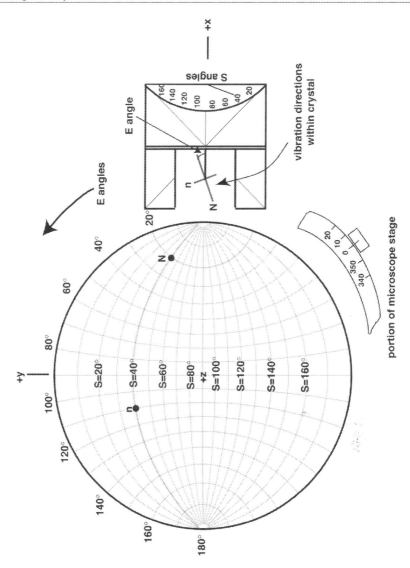

Fig. 9.4 Schematic showing the conventional directions for the axes of the Cartesian and spherical coordinate systems. The directions for the two coordinate systems are provided: 1. a Cartesian system (x, y, z) and 2. a polar system (S, E). The right-handed Cartesian system has the +x direction parallel to the spindle axis direction. The +y direction is perpendicular to x. Both x and y are within the plane of the microscope stage and the z+ direction is perpendicular to the projection/microscope stage, plotting in the middle. The S-angle is measured on the spindle stage dial as shown on the right. The S-angles are represented by great circles in the stereographic projection. The line representing the S = 40° plane is shown in bold. The S-angle defines a plane within the crystal that is parallel to the microscope stage. The E angle is measured on the microscope stage, and is represented by small circles on the stereographic projection. The line representing the E = 20° small circle is shown in bold. The E-angle is the angle between the spindle axis and a vibration direction in the crystal. (After Gunter et al. 2004)

Technical Text Box 9.1

PRINCIPLE OF STEREOGRAPHIC PROJECTION

The stereographic projection is a solid geometric technique of projecting 3-D objects on to a plane such that angular relationship between the objects is truly preserved. It is widely used in crystallographic and optical studies. Both crystal and optical parameters can be simultaneously plotted to yield a pictorial depiction of mutual relationship clearly. Understanding the construction of stereographic projection is a two step process, though today a number of PC plotting packages for stereographic projection are available.

The first step is to imagine that a crystal is enclosed in a sphere whose centre coincides with that of the crystal. The diameter of the crystal is arbitrary chosen but is big enough to completely enclose the crystal. In Fig. 1 the sphere and the crystal are shown; POP' is the vertical diameter; NOS and EOW are the horizontal diameters of the sphere. Draw from the centre poles to each face of the crystal such that these intersect the sphere as shown in Fig. 1. The logic of this construction is that since

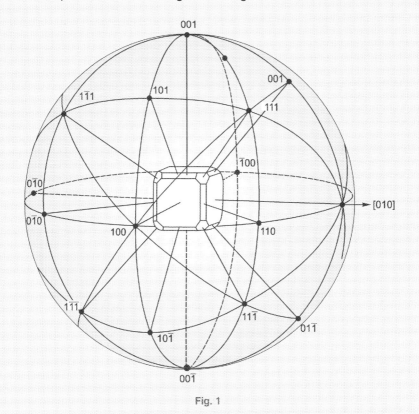

Fig. 1

Contd. ...

crystal faces have fixed relationship to each other for a given crystal, the points of intersection thus drawn are also unique to that crystal for a given diameter of the sphere. These points are known as spherical projection of the poles to crystal faces. Now, for the second step, join all points of intersection on the upper hemisphere surface with P' point of the vertical diameter POP'. These lines will intersect the horizontal plane defined by the circle NESW. Following our logic above we may state that points of intersection made on NESW also have definite relationship with each other that ought to reflect the angular relationship of our crystal faces. NESW is called primitive circle, primitive, or stereo-circle. The plot of points on the primitive is known as the stereogram of the crystal faces. Different optical parameters can also be plotted. Trignometric relationship is shown in Fig. 2.

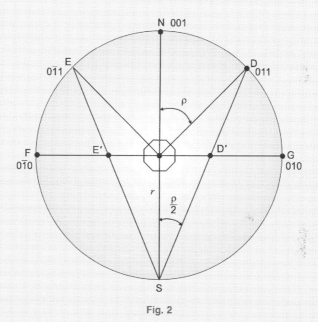

Fig. 2

Stereographic projection employs polar co-ordinates: r, ñ, ö. r is the distance of a given location from the origin, ñ is the angle, measured in the vertical plane between POP' and the desired vector of r units, and ö is the angle, measured in the horizontal plane, between the horizontal projection of the vector and a given reference line. The location of this reference line depends upon the objective. For example, in morphological crystallography, y-axis is reference lines and ö is measured clockwise as 0° to 360°, but ñ is expressed between 0° and 90° clockwise or anticlockwise from P. In the spindle stage and U-stage measurements, the ö-reference line depends upon the design of the equipments as shown in respective sections.

consists of a Wulf net plot and a table (Fig. 9.4) that give spindle stage orientations of optical parameters. For uniaxial crystals these parameters are optic axis, ù and å directions (two of these will coincide). For biaxial minerals the output gives value for 2V, and orientation for acute bisectrix (Bx), obtuse bisectix (OB), optic normal (â direction) and the orientation of the two optic axes (OA_1 and OA_2). After this you may reorient the crystal on the spindle stage again according to Bx values; use lambda plate to find out whether it is slow or fast direction; if fast, then the direction is á, and the mineral is biaxial negative. If Bx direction turns out to be the slow direction then the mineral would be biaxial positive (*See* Chapter 4, page YY).

After these parameters are determined for white light, the crystal may be mounted again and monochromatic light of more than one wave lengths may be employed one by one to measure the nature and amount of dispersion in anisotropic minerals.

You may have been wondering that not a word has been said about how the initial orientation of the grain is to be decided in the 'Bloss Method'. The grain should be mounted in a random direction, *i.e.,* neither parallel nor normal to any optic direction. To determine the "most favourable" direction, the initial positioning of the grain is tested by '40°' rule. The rule requires that after mounting the microscope stage be turned to 40° and the spindle stage is rotated slowly. At some point during rotation, the mineral will exhibit extinction; continue to rotate through 180°, you may notice another extinction. Repeat the procedure after rotating the microscope stage to 140° position. If the total number of extinction positions noticed by you during both 40° and 140° positions of the microscope stage are four, then this orientation is "most favourable"; three extinction positions are tolerable, but if the positions are less than three the grain should be repositioned till the four extinction positions during the 40° test are attained.

Universal Stage

The universal stage uses thin sections and yields quantitative data on 2V, extinction angle and also optic orientation. In case correlation charts between these parameters and the chemical composition are available with you, the universal stage provides an inexpensive, quick, and non-destructive method of obtaining compositions. However, microprobe techniques (electron- and ion) are now commonly available; at the same time are fairly accurate, though quite costly in terms of equipment time. But the utility of universal stage (often abbreviated as *U-stage*) for determining twin laws, orientation of exsolution lamellae, orientation of *c*-axis of quartz grains in tectonites, and a few other examples, is well-recognized.

In contrast to the spindle stage, the design of a U-stage is complicated. It is advised that you should get acquainted with all its components and accessories before an attempt is made to mount the U-stage on the microscope stage. While spindle stage can be used with any microscope, U-stage cannot be; hence, it is best purchased along with a given model of microscope. If it is being ordered separately, then it must be ensured that it would fit on one of the available microscopes. In view of its versatility in rotating the thin section (obviously the indicatrix as well) it is extremely important that the optic system of the microscope gets unified with that of the U-stage once the latter is mounted.

Design of the U-stage

The conspicuous feature about the design of the U-stage is its two storey fabrication. There is a base (Figs. 9.5 and 9.6)[4], A, that consists of a broad steel ring of such a diameter that it covers the microscope stage almost entirely; this ring has three threaded holes with screws that fit in the holes in the microscope stage for securely holding the U-stage. The centre of the U-stage, its *base ring*, A, exactly lies within the microscope axis. There are two raised arms attached to the base at the extremities of a diameter; normally these are placed as E-W

Fig. 9.5 A photograph of a Five-axis universal stage.

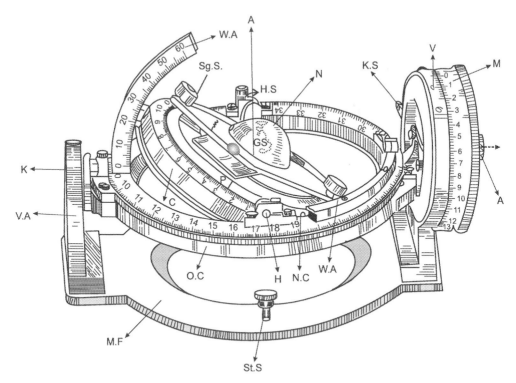

Fig. 9.6 A sketch of the design of a universal stage.

---

4. Readers are referred to two excellent monographs: Emmons and Naidu.

when the U-stage is properly mounted. The height of these arms is between 2 to 3 cms, and the arms act as hinges on which three concentric rings are attached, which can be rotated along E-W, N-S, and inclined axes to facilitate the viewing of the desired section of the grain. The rotation along any of these axes can be read on graduation engraved on the rings or attached devices (Fig. 9.7). When these readings are noted in the lab record,

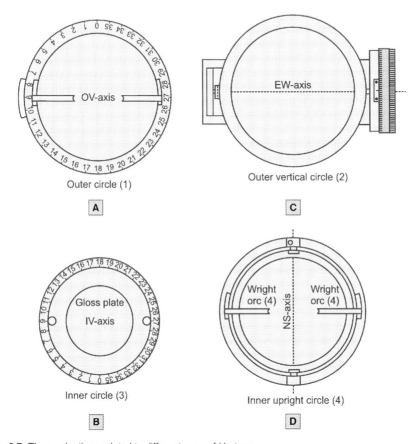

Fig. 9.7 The graduations related to different axes of U-stage.

a symbol is given to them that allows an easy tabular recording. Unfortunately, there has never been an attempt to standardize the nomenclature. Various well-known petrologists of early twentieth century used different notation that are summarized in Table 9.1. In the present work Reinhard's notation is followed.

It is important that one is familiar with the nature of graduation on various axes because it helps in data recording and subsequent plotting. The A-axis (Fig. 9.7A) is graduated from 0° to 360° anticlockwise such that the 180° marking is towards the observer and 270° is on the drum (K-axis) side. The N-axis (Fig. 9.7B) is engraved

Table 9.1 Notations used by some well known authors for different axes of U-stage

	Emmons	Reinhard	Berek	Dupare-Reinhard	Fedrov-Nitikin
Inner vertical (I.V)	N	A_1	N	N
Inner east-west (I.E-W)	–	–	–	–
North-south (N-S)	H	A_2	H	H
Outer vertical (O.V.)	A	A_3	M	M
Outer east-west (O.E-W)	K	A_4	J	J
Microscope Axis (M)	M	A_5	–	–

from 0 to 360° clockwise with 0 facing the observer; the readings are taken with reference to a pointer on the inner disc on which thin section is placed. The metal ring of the N-axis has two holes that allow a glass in the form of a hemisphere to be clamped by means of segment screws (Fig. 9.7). The K-axis readings (Fig. 9.7C), marked on a *drum* (placed on the right side of the observer), 0° to 360° markings, are in anticlockwise direction, *i.e.,* that the drum (fitted vertically) markings increase from the side facing the observer. For measurement of the tilt around H-axis, two metal arcs, named *Wright's arc*, are fitted (WA in Fig. 9.7D) on the metal ring of A-axis. These arcs can be raised and engravings on them (0 to 60°) enable one to record H-values. It must be clearly understood that the arc on which the tilt is measured is in the opposite direction of the tilt. When we depress the metal ring of N-axis on the right, the amount of tilt is read on the left arc. In order to avoid confusion our data table should also clearly indicate left or right arc along with the H-readings.

A mention has been made above about the position of a *glass hemisphere*. In fact, each U-stage comes with certain accessories; a set of hemispheres is such an accessory, and made of transparent material. Each set of hemispheres consists of two identical hemispheres so their combination becomes a sphere. The thin section is sandwiched

Fig.9.8 The upper and lower hemispheres of the universal

between the hemispheres of one set (Fig. 9.8). Note that although in shape the two hemispheres of a set are identical, the metal housing of each is different, *i.e.,* the top hemisphere housing is different from its bottom counterpart. The manufacturer supplies these sets since each set is characterized by a particular refractive index of the material of the hemisphere. The choice of a particular set for a given mineral to be studied is determined by the closeness of the refractive indices of the hemispheres and that mineral. The hemispheres allow one to view clearly thin sections under different tilt positions.

The U-stage is also supplied with special objectives because of the raised nature of the thin section location. It is necessary to withdraw the nosepiece containing usual objectives, and fit an assembly that will take the U-stage objectives; in addition to changed focal lengths the objective lenses are designed to receive imges from the curved surface of the upper hemispheres. Usually three objectives with differing focal lengths are sufficient; the highest one is suitable for oil immersion. These are marked as UM to distinguish them from the normal objectives of the microscope. It may also be necessary to use special condenser in place of the commonly provided condenser.

Adjustments of the U-stage

In order to fully utilize the capabilities of the U-stage, we must ensure that the optical system of the microscope is in unison with that of the U-stage. The first point to watch is whether N-axis of the stage coincides with that of the microscope in the zero position. It is advisable that this adjustment is done before the thin section is mounted with the hemisphere. This is noticed when an inclusion in a grain of the thin section does not remain stationary during the rotation of N-axis. The eccentricity is removed by adjusting the U-stage slightly with the help of base screws. The base screws are loosened slightly; then N-axis is rotated, and when the marked inclusion is farthest from the centre of the field of view, the U-stage, as a whole, is shifted manually and very carefully by utilizing the small play that got into existence because of the loose base screws, so as to bring the inclusion half way closer to the centre. Then the thin section is moved in such a way that the marked inclusion is back at the centre. Now, the N-axis is again rotated and it is watched whether the marked inclusion now remains stationary. If yes, then the first adjustment is over, otherwise the above steps are repeated till the perfect centricity is obtained. You will notice that this procedure is similar to that of removal of eccentricity of objective lenses (Chapter 5, page YY). Once centricity is achieved the U-stage screws are firmly tightened. N-axis is returned to zero position.

Next, we have to make sure the H- and K-axes are parallel to the crosswires of the ocular in the zero position of the U-stage. In order to test this alignment, the hemisphere is clamped on the metal ring of N-axis, and a dust particle on the top of the hemisphere is focused by means of a medium power UM objective. When K-axis is rotated (both clockwise and anticlockwise) the particle should move along the N-S crosswire. If instead of moving along the crosswire the particle moves across it, then an adjustment is required. The lock of A-axis, normally located at the centre of the drum (K-axis) is loosened; a slight manipulation of A-axis (usually <1°) either way corrects the misalignment; the lock is then clamped again and new reading of A-axis (original reading was 90°, *see* description above, also Fig. 9.6) is noted as 'zero reading'. This exercise is repeated for movement along E-W crosswire by tilting along H-axis, and aligning the K-axis along E-W by a small manipulation of A-axis if required.

Finally, we have to ensure that both H- and K-axes are within the plane of the mineral grain of the thin section. Though manufacturers take great precaution in the design of the stage, yet it is possible that the grain section is slightly above or below the plane containing the H- and K-axes. To ensure this coincidence, threads are provided

Reorienting Techniques

below the metal ring of N-axis along which the thin section can be raised or lowered by a small amount. This movement is carried out by holding one of the four levers (called 'handles') attached to metal ring. A small clockwise rotation raises the thin section and anticlockwise rotation depresses the thin section. To test for the correct height of the metal ring, a grain is brought to the intersection of the crosswires and K-axis is rotated. If the height is correct then the grain remains stationary, otherwise it would appear either raised or depressed. In that case a slight rotation of threads by means of 'handles' provided below the metal ring, as mentioned above, would bring the proper alignment.

Fundamental Principle

The U-stage tries to bring the optic indicatrix (Chapter 4) alignment in favourable position for viewing. Once the orientation and nature of indicatrix is known, then, with suitable measurement, and subsequent manipulations on a stereogram, useful information can be extracted easily. It is primarily because of this reason that the stereogram, popularly known as *Wulf net* (Technical Text Box 9.2), is graduated exactly like N-axis. Therefore, the first step is to bring the mineral to extinction position, thereby bringing trace of optic directions of the mineral parallel to the vibration directions of the nicols. The optic plane parallel to N-S crosswire is normal to K-axis. This optic plane forms a great circle on the stereonet and its pole will lie at the extremities of K-axis, *i.e.,* at 90° to the great circle. In general, we shall require manipulations on N-axis (by bring the grain to extinction), and then tilt around H-axis to make a plane vertical.

Uniaxial Crystals

In Chapter 4, the uniaxial indicatrix has been described and you are advised to refresh the material given therein for a clear understanding of this section. The most important property in a uniaxial crystal is the optic axis which coincides with its *c*-axis (Page YY). You also know that a ray propagating along the optic axis would be stopped by the analyzer resulting in the complete extinction of view. By chance, it may so happen that you would be able to find a grain of a uniaxial mineral that remains extinct throughout the rotation of the N-axis without any manipulation of the U-stage axis; then, this grain has been sectioned normal to *c*-axis. But, in general, a grain of the uniaxial mineral will show polarization colours under cross nicols. Therefore, it is either an oblique section (Case A), or *c*-axis parallel (Case B). We shall examine each case separately.

Case A: Oblique Section

Fig. 9.9A shows a stereogram where point O represents the optic axis in a given oblique section, and, say, plots in the third quadrant in this example. In order to ascertain the location of the indicatrix with reference to the plane of the thin section, we bring the grain to its extinction position by maneuvering N-axis. You know that in this position

> **Technical Text Box 9.2**

WULF NET AND ITS APPLICATIONS

The Wulf net (Fig. 1) is provided in this book for your convenience which can be scanned and stored in your personal computer. The outline of the net is same as that of the primitive circle; the general experience of a beginner is that a circle of 10cm

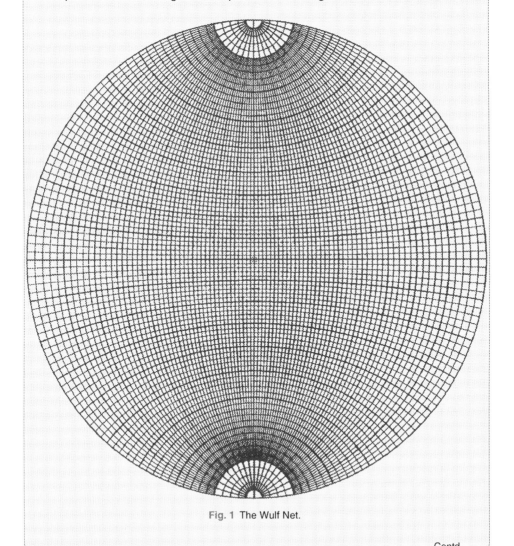

Fig. 1 The Wulf Net.

Contd. ...

radius is good enough, though some prefer a circle of 5cm radius. The net design consists of a series of circular arcs of two orientations. Arcs of one series are drawn between N-S points. Since N-S is diameter these arcs become great circle (*see* Technical Text Box 9.1). With N as the centre concentric arcs are drawn as the second set such that E-S diameter would also represent one such arc. Similarly, as a part of the second set concentric arcs are drawn with S as the centre till an arc coincides with E-S diameter. These concentric arcs are called small circles. Now our primitive circle acquires a design of a net, hence the name.

You may notice that both N-S and E-W diameters, although appear as straight lines, are also great circles, so are the semi-circumference between N and S. The semicircular space within the semicircle defined with NS as the chord is divided into 90°, and the great circles are drawn at 2° interval. Every tenth great circle, *i.e.*, 10°, 20°, and so on is marked by a bold line. Likewise, the small circles are also drawn at 2° interval, and every tenth small circle is in bold line.

For manual practice, it is always preferred that you use a tracing overlay on the given Wulf net. The size of the overlay is always larger and you use extra space to write date and aim of the experiment plus any other information that could matter for future record.

First a table is generated in the lab record book of ñ and ö values for each set of observations. The value of ö is counted from the reference point on the periphery of the primitive circle; ñ values are plotted with reference to corresponding ö values inside the primitive circle but the manner of plotting would differ from one objective to the other. Some examples are given in the relevant sections; the students may also consult standard textbooks on crystallography for details.

the trace of the principal section of the uniaxial mineral is parallel to the vibration direction of one of the polars, though it is immediately not known whether the vibration direction of the lower nicol (corresponding to N-S crosswire) is parallel to the trace of the principal section of the mineral; it could be, alternatively, parallel to that of the upper nicol (corresponding to the E-W crosswire). Our aim is to make the principal section of the mineral coincident with N-S crosswire,

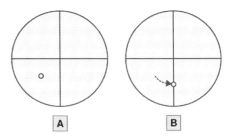

Fig. 9.9 The stereonet plot of an oblique section of a uniaxial crystal.

and then employ K-axis to bring optic axis of the mineral in vertical position; in this position the optic axis would coincide with the microscope axis and complete darkness would result during the rotation of the microscope stage. This situation is depicted in Fig. 9.9B. At this point we shall record the rotation achieved on the inner circle (N-reading in degrees: 0° to 360°).

The first step is to explore for the possibilities of the principal section being aligned parallel to E-W. For this, a tilt around H axis is made, if, one notices improvement of extinction (grain becomes darker), and at one position it is at its darkest nature, then the principal section is parallel to E-W; so we rotate N-axis by 90° in order to bring the principal section along N-S. If, on the other hand, the manipulation around H-axis brings illumination, then the principal section is already N-S, so we take the next step of achieving extinction by rotation of K-axis. After the grain becomes extinct by rotation along K-axis, rotate the microscope stage by a few degrees. It is quite possible that grain gains illumination, so manipulate K-axis to bring back the extinction, and, then rotate the microscope stage again by a few degrees to see if the extinction persists. If not, then use K-axis to bring the grain again to extinction position. This happens because your initial step of bringing coincidence of the optic axis of the grain with microscope axis may not result in the exact coincidence in the very first step itself. A careful repetition of these exercises will achieve the coincidence. This is tested by carrying out rotation around microscope axis and observing that extinction persists throughout the rotation. The N and K readings are noted for deciphering the orientation of *c*-axis of the uniaxial grain. These are respectively equivalent to ö and ñ as given Technical Text Box 9.2. If the section of the grain can be related to the geographic directions of the location of the outcrop from where sample has been picked, then the optic axis tilt with respect to the ground can be easily noted.

Case B: *c*-Axis Parallel Section

When a section of a grain is cut in such a manner that *c*-axis lies within the plane, or it is very nearly parallel to the section, then, the above method will not yield the result. In this case bring the grain to 45° position after obtaining extinction. The grain

Reorienting Techniques 167

will be illuminated; rotate along the K-axis first around clockwise and then around anticlockwise directions. If the colours rise in both rotations then the optic axis is along E-W directions. If the colours fall in both directions then the optic axis is along N-S direction. This is easily conformed by first bringing microscope axis to its initial position then rotating the grain on N-axis by 90° (this should bring the optic axis now parallel to N-S direction). Now, bring the grain to 45° position and rotate K-axis first clockwise and then anticlockwise. The interference colors should rise during the rotation of K-axis.

Optical Properties

Having determined the location of the c-axis in the given grain it is fairly easy to determine important optical properties quite accurately. Let the axes of the U-stage be in the initial position and the desired grain is in the centre of the field of view. Bring the optic axis in the horizontal plane by rotation and tilting on suitable axis as already determined. Now, introduce the ë or ë/4 to determine whether the optic axis represents fast direction or slow one. We know from Chapter 4, page YY, that a fast optic axis direction means a uniaxial negative crystal; likewise, a slow optic axis direction indicates uniaxial positive crystal. If the grain is elongated (most uniaxial minerals are so by habit) fast direction coinciding with the length (c-axis) would mean positive elongation and uniaxial negative crystal. Slow direction coinciding with the length (c-axis) would mean negative elongation and uniaxial positive crystal.

If the mineral is pleochroic then the pleochroic scheme is now easy to determine. Take the upper polar out when the optic axis is along the N-S direction within the horizontal plane. As you know from Chapter 4, page YY, N-S direction is the vibration direction of the lower polar, hence, the colour of the mineral in this position is the colour shown by å. Rotate the N-axis by 90° the colour of the mineral, now, is its colour shown by ù.

Biaxial Crystals

By now it is obvious to you that the first step with the universal stage is to bring the grain to its extinction position irrespective of the fact whether it is a uniaxial mineral or biaxial one. There are three optic planes in biaxial minerals instead of two for the uniaxial minerals, therefore, a biaxial grain in the extinction position has traces of two of its optical planes along the crosswires. The problem is to find out which two of the three optic directions (á-â, á-ã, or â-ã) are being viewed? It is obvious that a complete description of optic properties of a biaxial mineral would require two grains such that between these two grains all three optic directions are available. In other words, if one grain section present á and â, then, the second section would be an á-ã section. The á-ã section, being the optic axial plane (OAP) is always one of the most sought after section.

The first step is to make one of the optic planes vertical, and this is achieved by testing whether during the rotation of K-axis the extinction persists; if so, then the plane is vertical, otherwise it is not. The plane could be inclined towards right or left, therefore, a tilt around H-axis in the opposite direction will make the plane vertical; this is easily noticed as the extinction would persist with the rotation around K-axis. It may also require a slight adjustment of N-axis as well.

Once the above step is over, H- and N-axes are left undisturbed in their position, and the microscope stage is rotated by 45° bringing the grain to its brightest position. In this situation the optic plane is now in 45° position; the two optic directions (á, â, or ã) that lie within it (*i.e.,* the constituent optic directions) are inclined to the microscope axis[5]. If this optic plane is á-ã, *i.e.,* optic axial plane (OAP), then it also contains the two optic axes. One or both optic axes can be made to coincide with the microscope axis. This is tested by rotating K-axis both clockwise and anti-clockwise. As the rotation begins, the order of the interference colours would change; it may decrease to extinction; at this point one of the optic axes would coincide with the microscope stage. More rotation in the same sense or opposite sense may also bring out the second optic axes, particularly if 2V is small. Optic direction â is now normal to microscope axis. You must record the rotation of K axis on the vernier provided.

Alternatively, the extinction may be achieved by either â-ã or á-â plane but not á-ã plane, and then it is observed that once the optic plane is brought to the vertical position, the rotation around K-axis does not decrease the interference colors to black. In order to test whether it is a â-ã or á-â plane we make use of tinted plates, say ë plate. Bring the microscope stage to its 45° position. With ë plate in the slot, if the lowering of the order of interference color takes place then the slow direction of the plate is coinciding with the fast direction of the grain *i.e.,* the grain direction would be á and the plane we are investigating, therefore, would be â-ã.

After we have located one optic plane we must now proceed to locate the second plane, but before doing so make sure that data for the first plane, *i.e.,* N, H, and K readings are properly recorded in the lab record. Return the microscope and U-stage to their initial positions. A rotation by 90° around N-axis from its previous position would bring the grain, again, to its extinction position. A repeat of steps carried out to locate the first optic plane is done including the final testing by ë plate. All the readings are carefully recorded. Since the third optic plane is normal to these two, its location is ascertained by geometric exercises being described in the following pages.

It may be useful to record the orientation of cleavage planes and/or twin planes of grains. We know the relationship (Miller indices) of a cleavage plane or twin plane to the crystallographic axes from textbooks. For example, muscovite has a basal cleavage,

---

5. In a chance thin section of the grain, one of these directions may actually coincide with the microscope axis.

amphiboles and pyroxenes have prismatic cleavage, or albite twin plane is the side pinacoid. In several instances, a prominent pinacoid trace is easily recognized. Such features guide us to locate the crystallographic axes of the grain. The orientation of such a crystal plane with reference to the section of the grain under view is easily determined by bringing the trace of the plane parallel to N-S crosswire through manipulation of the N-axis. As in the previous cases, though the trace is parallel to N-S but the plane may or may not be vertical. Now a careful and slow tilt around H-axis will bring the plane in the vertical position which will be evident from the extreme sharpness of the trace; the trace will become focused. Most experienced workers suggest that the iris diaphragms below the polarizer and the objectives should be partly closed with approximately similar diameter when this exercise is taken up. With the plane now in vertical position, the pole of the plane becomes coincident with K-axis. N and H axes readings are noted for this position.

Optical Properties

A method has been outlined, in the preceding section, which determines the fundamental optical properties of a biaxial mineral, *i.e.*, the three optic directions and 2V. You already know that there are five possible sections in which a grain of a biaxial mineral may be preserved in a thin section (Chapter 8, page YY). These five possibilities combine with a wide range of values of optical data such refractive index, birefringence, 2V etc., present challenges in determining optical properties, including, at times, even identification of a given grain. Hence, you will need a strong theoretical background of the principles outlined in the preceding pages as well as imagination to solve problems.

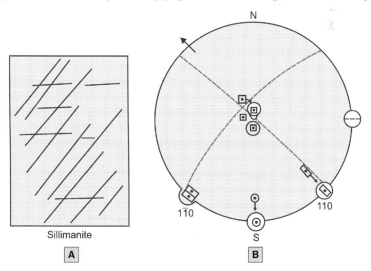

Fig. 9.10 A. A euhedral grain of sillimanite; **B.** Its optical parameters as plotted on the Wulf's net.

---

6. An example from Naidu (1960).

Table 9.2 U-stage data on sillimanite

Parameters	N	H	
â direction	142°	11°	
Cleavage	142°	11°	
Short prismatic side	7°	3°	
Long prismatic side	96°	18°	
Optic axes on K-axis	$K_1 = 356°$ $K_2 = 335°$		
∴ 2V	21°;		ë-plate gives ã as acute bisectrix

As an example, we take sillimanite gneiss from Madurai, Tamilnadu[6]. In this example the grain of sillimanite chosen is nearly euhedral such that (010) cleavage is well seen and the grain outline is nearly rectangular (Fig. 9.10A). The rectangular sides are traces of prism faces. The U-stage observations yield the data given in Table 9.2.

Prof. Naidu's long experience helped him in the grain selection. It is advisable that you attempt to develop the ability to recognize which one of the five possible orientations you are looking at. It should be possible to gain such an expertise given a certain amount of practice. The advantage of this selected grain is that it allows measurement of 2V directly; hence, we do not need additional measurement. A stereoplot of the data in the above table is given in Fig. 9.10B. The steps for plotting are simple, *i.e.,* first locate â on the stereonet. Then bring this pole on the E-W line and count 90 from it; the great circle at this point is traced and would represent a projection of á-ã plane which, by definition, is the Optic Axial Plane (OAP). Plot the two optic axes *along this great circle* (K_1 and K_2 readings). The mid-point of these two optic axes *along this great circle* is ã direction; count 90° from ã and mark á. Now, it must be evident to you that fewer observations ensure fewer errors as well as savings on time. This is particularly true when the mineral is coloured, consequently, the absorption is high.

We may now proceed for obtaining optic orientation of the grain (Chapter 4, page YY). Bring ã direction pole to the centre of the stereonet by counting the degrees on the small circle. Mark it as ã with a different coloured pencil as ã´. Transfer all points on the plot along the respective small circles by identical amounts along the same direction. This will bring á´, new plot of á, on the periphery of the stereonet. Rotate the transparent overlay till á´ faces the observer. You will also notice that the rotation exercise has not only brought á´ on the periphery, but also the poles of the prisms as well as cleavage are also now on this periphery. Note that á´ bisect the angel between the poles of the prisms, therefore, coincides with a crystal axis. (010) cleavage pole is at 90° to á´ and coincides with â´ therefore, â´ is along y-axis of the crystal, á´ becomes coincident with x-axis giving us the complete optic orientation.

Although sillimanite is a colourless mineral, but similar observations on a coloured (pleochroic) mineral will also help u s to write down pleochroic scheme and the absorption formula once the optic orientation becomes known.

You may have come across a well-known crucial question: whether a given grain belongs to pyroxene or amphibole group? Petrologically both could be possible for a given thin section; however, no grain is present that exhibit the two prismatic cleavages. You have the option of taking a probe shot at it, but it requires a delicate thin section preparation, and, of course, the EPMA handy (not a easily met condition!). The optic orientation method outline above will tell without additional expenditure of an EPMA. The data when plotted on the stereonet (such as in Fig. 9.10) will yield the pole positions of the single cleavage and á. Read the angle between them along the great circle. If the angle is 45° then it is a pyroxene; if the angle is 60° then it is an amphibole.

Optic orientation studies in Monoclinic System present problems in the sense that only one of the optic direction coincides with y-axis. In the case of pyroxenes and amphiboles it would certainly help to get a grain that exhibits both cleavages. When only one cleavage is present, particularly in minerals like epidote where only one set of cleavage is possible, we must rely on additional criteria. It is a common practice to transfer â direction pole to the centre of the stereonet.

The U-stage procedure for triclinic minerals, like microcline, is a bit longer since none of the optic directions coincides with any crystal axes. We must choose at least two grains with different orientations; cleavages, when present, would guide us in such situations. The data from both grains are plotted on the same stereonet with a care that labels should be clear and distinctive. Transfer â of both grains to the centre. We take another example of Madurai District, this time of wollastonite. The data is as given on Table 9.3.

Fig. 9.11 depicts stereonet projection of the given data (Table 9.3) now rotated with â in the centre. With â at the centre note that á and ã of both grains coincide with respective directions. In this situation note that cleavage poles of both grains now lie on one zone with zone axis nearly coinciding with â. From textbooks we know that in

Optical Mineralogy

Table 9.3 U-stage data on wollastonite

I. Parameters	N	H
â direction	52	14°
Cleavage	321°	26°
	152°	22°
Optic axes on K-axis	$K_1 = 349°$	
	$K_2 = 28°$	
∴ 2V	39°;	
II. Parameters	N	H
² direction	47°	37°
Cleavage	144°	6°
	237°	35°
	179°	40°
Optic axes on K-axis	$K_1 = 351°$	
	$K_2 = 307°$	
	44°	

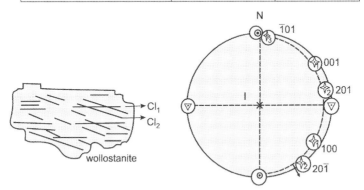

1,2 and 3 = Cleavages from three different grains.

Fig. 9.11 A grain of wollastonite and its optical parameters as plotted on the Wulf's net.

wollastonite, cleavage planes lie in a zone whose axis is y-axis. This allows us to plot y-axis and we can measure y∧â as well.

Reorienting Techniques

Summary

To measure accurately refractive indices of anisotropic minerals, to obtain the optic indicatrix, and to ascertain the relationship between crystallographic axes and the optic directions, rotation devices have been designed. Spindle stage and Universal stage are two common rotation devices.

The spindle stage consists of a metal wire formed in the shape of a spindle attached to a graduated semi-circular disk. The disk is utilized to note the amount of rotation of the spindle. Normally, a euhedral or subhedral (with sufficient face development) grain is chosen and its optic directions already as mounted on the stage. Then one by one each direction is brought parallel to the spindle axis, dipped in the appropriate R.I. liquid; by following the Becke line method ù and â of uniaxial minerals, and á, â, and ã of biaxial minerals are determined. Bloss method employs mounting a crystal in random orientation (after 40° test method) and then plotting a succession of extinction angle. With the help of a computer programme, EXCALIBRW, the spindle stage orientation of all optical parameters are calculated. These are used to determine refractive indices and dispersion.

U-stage employs normal thin sections. There are 4 or 5 axes of rotation in a U-stage; two of these are horizontal and at least two are vertical in the initial setting. The stage requires special objectives and a series of adjustment tests are carried out to ensure perfect centricity of level coincidence. The fundamental principle is to manipulate rotation amongst axes of the stage in such a way that one of the optic directions of the mineral coincides with the microscope (vertical) axis.

The list of U-stage capabilities is very long, but we shall rest here. The reader is referred to the two excellent books on this subject upon which the author has always relied. A number of charts are given in these references that help in various determinations including twin laws of feldspars etc.

Important Terms

40° test for spindle stage 158

Base ring 159

E and S values 153

Fedrov as a 4-axes stage 150

Glass hemisphere 161

height adjustment in U-stage 160

H and N axes 168

rotation 150

spindle stage 150

U-stage 153

universal stage 150

Wilcox's spindle stage 151

Wright's arc 161

Wulf net 154

Questions

1. What is the purpose of employing rotation techniques?
2. How many optical parameters can be measured by these techniques?
3. Describe the construction of Wilcox's Spindle stage.
4. What is the main application of the spindle stage?
5. How would you orient subhedral uniaxial grains on the spindle stage?
6. What is the 40° test for mounting grains on a spindle stage?
7. Give a brief description of the input data for EXCLIBRW.
8. How would you determine dispersion of a crystal by means of spindle stage?
9. What is the characteristic architecture of U-stage?
10. How the U-stage is mounted on the microscope?
11. Describe the graduation on scales attached to various axes.
12. What purpose do the hemispheres serve in the observations?
13. Describe the assembly of the hemispheres.
14. What is the need for centering N-axis? How is it done?
15. What is the need for height adjustment? How is it done?
16. How would you determine the orientation of c-axis of numerous quartz grains in a quartzite thin section?
17. For 2V determination of a fine grain mineral which stage: spindle or universal, is more suitable? And why?
18. Describe briefly the difficulties one would face in determining optical parameters in monoclinic and triclinic crystals.

Suggestion for Further Reading

Bloss, F.D., 1981, The Spindle Stage: Principles and Practice, Cambridge University Press: Cambridge, England.

Bloss, F.D., 1999, Optical Crystallography, Mineralogical Society of America.

Emmons, R.C., 1943, (Reprinted 1964), The Universal Stage, Geol. Soc. America Mem., 8, 205p.

Gunter, M.E., B.R. Bandli, F.D. Bloss, S.H. Evans, S. Su and R. Weaver, 2004, Results from a McCrone Spindle Stage Short Course, a New Version of EXCALIBR, and How to Build a Spindle Stage, MICROSCOPE, Vol. 52:1, 23-39.

Naidu, P.R.J., 1960, Four Axis Universal Stage, Private circulation.

10
New Frontiers in Microscopy

Learning Objectives

- Secrets of fluid inclusions
- Interior of the mineral structures
- Higher resolution graphics of grains in thin section
- Surface textures of minerals and rocks.

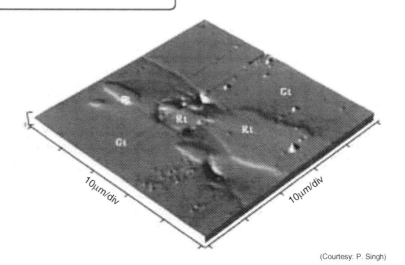

(Courtesy: P. Singh)

An AFM image of a part of garnet grain with rutile inclusions.

Prologue

The vision must be followed by the venture. It is not enough to stare up the steps we must step up the stairs.

— Vance Havne

Chapter's Outline

- Fluid inclusion studies
- Image analyzing system
- Transmission electron microscopy
- Scanning electron microscopy
- Atomic force microscopy

© The Author(s) 2023
P. K. Verma, *Optical Mineralogy*,
https://doi.org/10.1007/978-3-031-40765-9_10

Introduction

This chapter deals with techniques that are not directly related to identification of minerals under microscope but for specialized investigation employ principles of optics in one way or another. Of course, techniques like fluid inclusion studies are directly related to microscopy with an accessory stage mounted on the microscope.

Fluid Inclusion Studies

In some of the preceding chapters we have tried to impress upon you about the kind of information a mineral grain has stored for us; the information that reveals the environment that was present when the rock was forming, and about the subsequent events that affected the rock in any minor or major manner. In Chapter 6, where systematic examination of optical properties is described, Table 6.1 has a data entry called 'Inclusions'. Initially, you were told that a large grain called, host grain, may contain small grains of other minerals, rarely the same mineral, which are called inclusions. There are two other types of inclusions that may occur in the host grain. One type is *melt inclusions*, which are tiny grains of the melt from which the mineral has crystallized. The melt inclusions now occur as glass shards or globules. These are studied by equipment called *Ion Microprobe*. The description of ion microprobe and techniques employed in studies on melt inclusions are beyond the scope of the present book. The second category is *fluid inclusions*, which are tiny cavities within the host grain, often containing a fluid phase; one or more solid phases may also be present within a particular inclusion. It is assumed that the fluid and the solid phases were part of circulating fluids, either of meteoric or of juvenile origin that played a major role in the deposition of the host grain. Because of this belief many petrologists pay special attention to these inclusions. Thin section preparation is slightly different as already mentioned in Chapter 2; it is considerably thicker (~100ì m) than the normal thin section (Fig. 10.1) and is very well polished. A special assembly of microscope and accessories are utilized for the fluid inclusion studies as

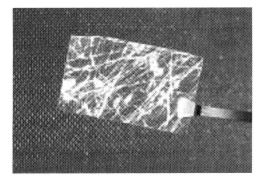

Fig. 10.1 A thick thin section for microthermo-metric studies.

shown in Fig. 10.2. On the extreme right is a polarizing microscope (a research model: Chapter 5, page YY) with a camera attachment at the top above the ocular assembly. The cables from the camera are connected to the personal computer set-up kept at the extreme left. On the top of the microscope stage an additional stage assembly is placed which is connected to a set of control panels that are commonly kept between the microscope and the pc for controlling temperature of the additional stage. The thin

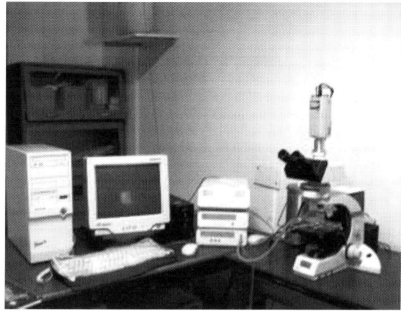

(Courtsey: A Saikia)

Fig. 10.2 A Microthermometry set up at the Geology Department, DU.

section containing the fluid inclusions is mounted on this accessory stage, also known as *heating-freezing stage* or, *microthermometric stage*. The sample holder for the thin section is a rectangular box with a glass window. Inside the box there is a thermocouple, a heating wire and an arrangement for the inflow and outflow of liquid nitrogen as is easily seen in Fig. 10.2. The working temperature range is from below 0°C, achieved with the help of liquid nitrogen[1], to over 600°C. The investigator watches through the ocular the physico-chemical changes taking place within the inclusion as the temperature is varied over the wide range mentioned above. These changes are recorded with the help of devices such as 3-D chip video camera. In several heating-freezing stages there is provision to record IR spectrum using CCD sensitive up to 1100nm. Some common interpretations in the microthermometric investigations yield density of phases and bulk salinity of the inclusions.

Image Analysis System

In addition to data on the optical properties, some physical parameters of mineral grains become important in several petrological, sedimentological, and metallurgical problems.

---

1. Liquid nitrogen is commercially available in vacuum flasks of different sizes, although many laboratories also have distillation plants where liquid nitrogen is fractionated from liquid air. Liquid nitrogen (LN_2) boils at 77°K, *i.e.*, – 196°C, and has the property of cooling the object very rapidly, though it is suggested that for real efficient cooling chunks of solid nitrogen mixed with liquid nitrogen are better.

For example, we may need to know how many grains are spherical or elliptical in a given thin section. To obtain quantitative data on geometric, volumetric and densitometric characters of grains is the purpose of *image analysis*. Before the image analysis became popular, mineralogists and petrologists were using various accessories to manually carry out these measurements. Some of these are still in use because of cost factor. Some examples are *camera lucida, mechanical stage* and *point counter, ocular graphs* etc. There was subjectivity in the data generation besides being very strenuous and enormously time consuming.

Mechanical Stage

Mechanical stage is an accessory stage, in fact, the name stage is misnomer to it because it consists barely of two orthogonal arms with two screws projecting downwards. The size of the two screws and the distance between them is such that these fit in the threaded holes provided in the microscope stage[2]. Once the mechanical stage is firmly placed on the microscope stage then a thin section may be placed on the latter in the usual manner except that the two spring clippers provided to the mechanical stage (Fig. 10.3) are manipulated to hold the section securely. The arms of the accessory stage now serves as x and y directions along which the thin section is moved by means of screws provided at an end of each arm as shown in Fig. 10.3. Movement by means of the screws advances the thin section by one unit in either x or y direction. This enables you to view the entire thin section at a controlled interval. The fact that the description of the mechanical stage is being given here does not mean that the accessory represents new frontiers in microscopy! On the contrary it has been in use since the beginning of the 20th century for a variety of petrographic investigations. It makes several exercises on the image analysis system easy and regulated.

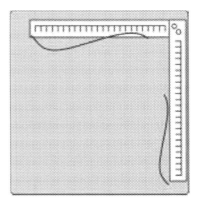

Fig. 10.3 A sketch of mechanical stage used for holding a thin section.

Image Analysis in Microscopy

Towards the end of the last century, application of high-speed personal computer brought in the concept of image analysis in microscopy. Fig. 10.4 is a photograph of an image analysis system supplied by Leica Co. which essentially consists of a high end research polarizing microscope with provision for switching from transparent beam to reflecting beam microscopy. The binocular ocular system has an accessory attached to it for

---

2. This coincidence may not happen if the manufacturers of the microscope and the mechanical stage are not the same.

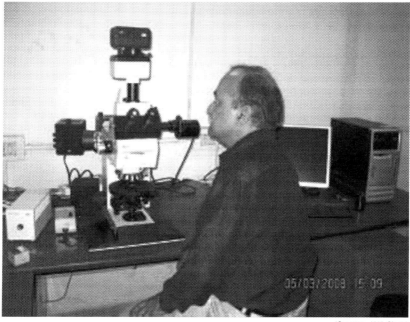

(Courtesy: K.K. Singh)

Fig. 10.4 An assembly of Image analyzer at Geology Department, DU.

inserting scale that gives the estimation of the enlargement. A digital camera is attached to the microscope tube, and a cable from the camera transmits signals to a nearby personal computer. Since normally the working conditions do not require high or freezing temperatures, there is no control panel as shown in Fig. 10.2. However, the microscope has additional features to help us investigate minerals and rocks better. All the rules for the good maintenance of a polarizing microscope given in Chapter 5 (pages YY) are also true for this image analyzer; in addition, camera and the computers must also be treated with tenderness and due care. Instructions given in the manual supplied by the company should be faithfully followed as a good working habit.[3]

Applications of Image Analyzer

Example 1. Modal Analysis

The first one is the modal analysis of a rock. The importance of the *modal analysis* of a rock may not be apparent to a beginner in geology, metallurgy, or material science, but, in a large variety of cases it is important to find out percentage by volume or weight of

---

3. After a bitter experience of past five years, the Geology Department of Delhi University has banned the use of external drives, including memory sticks etc. The users are now required to carry their data in CD's or DVD's. The author strongly supports this, though discs do pose a problem of storage

different major minerals or constituents. As an example suppose we are interested to know the progress of a reaction that produces epidote in an impure carbonate protolithic. Therefore, it may be important to know how, and, in what direction the percentage of epidote is increasing. One such rock is illustrated in Fig. 10.5. This is a relatively easy case since in this view abundant epidote grains are conspicuous because of their optical behaviour. Our object is to determine the volume percentage of major minerals in this rock, *i.e.*, quartz, plagioclase, epidote and calc-amphibole (not seen in the figure).

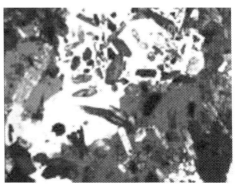

(Courtesy: Department of Geology Collection, DU)

Fig. 10.5 Photomicrograph of a calc silicate rock. Abundant nearly euhedral grains of epidote are conspicuous. Crossed polar view.

Step 1: It is advisable to demarcate sectors in the thin section so that each sector fills up almost the entire field of view of the microscope. As you know, it will depend upon the settings of the objective and the ocular.

Step 2: Now we open the software package for the image analysis. Manufactures may have different names for this package; the one in use by Leica Co. is called *Q Win*. There is command called *Live* that shows the field of view of the microscope on the monitor.

Step 3: With the help of command, *Calibrate*, we choose scale and enter magnification of objective and any other desired information.

Step 4: Then, the step to choose the grains is taken. In the present example, suppose epidote is first chosen, and then the cursor is brought over an epidote grain.

Step 5: By the help of command, *Detect*, the computer is asked to identify it. You will notice that it picks up all epidote grains under view and gives them a color, say, blue.

Step 6: It functions on the basis of optics; it might pick up non-epidote portions of the field of view, and may leave some epidote grains. There are corrective steps for this, which, finally, should result in perfect or near perfect selection. These steps are easier said than done and would require practice sessions.

Step 7: Go to the command *Measure*, which would ask for the required parameters; when this questionnaire is over, just press Enter. The tabular data would pop-up (Table 10.1). It indicates that within the field of view epidote grains occupy 58.44% of the surface area. This procedure is repeated for other minerals, then the either X or Y arm of the mechanical stage is moved one graduation mark or the next sector of the grid. This exercise continues till the entire thin section is covered. You may save each

Table 10.1 Binary Data from Image Analyzer

Field #	Area	Intercept H	Intercept V	Perimeter	Count	Area Fract	Area Fill	Area%	Count/Area	Meas.Frame
1	12292105	162273.4	196295.6	570193	4073	0.58	1.41	58.44	193.63	21034574

frame in a spread sheet and then find the total to get the overall surface percentage of major minerals. The data is available, now, for further evaluation.

Example 2. Grain Size Measurements

Grain size measurement of phenocrysts, porphyroblasts, or sand grains in thin sections is a revealing and important exercise. Fig. 10.6 shows a large garnet grain in an alaskite thin section. Such phenocrysts are abundant in this rock, and their *size distribution* is an important petrographic indicator. The problem here is that the garnet is zoned, plus portions of matrix are also detected along with the garnet grain because of similarity in some optical properties.

(Courtesy: Department of Geology Collection, DU)

Fig. 10.6 Photomicrograph of a garnet porphyroblast occurring in alaskite. Plane polarized view.

Step 1: Initial steps are similar to those described above. For this exercise we begin by demarcating the grain boundary. It is done with the help of the cursor that is moved solely along the boundary of the grain; since the boundary is irregular, the cursor movement ought to be carried out carefully.

Step 2: Demarcation of the grain boundary allows QWin to select the grain and impart a color of your choice. The false color may need improvement within the selected area, particularly inclusions within the grain and weird embayments in the grain boundary have to be manually sorted out. This method has been described in the first example.

Step 3: Now use measure command of QWin package; answer all the questions asked, and press Enter. It would yield the desired data. For the garnet shown in Fig. 10.6 the data is presented in Table 10.2. It shows that along X direction the grain has a length of 9958.82ì m or 9.95882mm; the length along Y direction is 9689.42ì m or 9.68942mm.

Table 10.2 Binary Data from the Image Analyzer

Field #	Area	Intercept H	Intercept V	Perimeter	Count	Area Fract	Area Fill	Area%	Count/Area	Meas.Frame
1	3787589	9958.82	9689.42	30482.61	25	0.25	0.33	24.93	1.65	15194101

Step 4: The procedure is repeated on another grain till sufficient grains of garnet have been measured in a given thin section. From this data set, statistical mean etc. can be easily determined.

Transmission Electron Microscope (TEM)

Transmission Electron Microscope or, TEM as it is popularly known, has been used for mineralogic studies for the past several decades. An electron beam is considered here as a wave with a very short wavelength (<0.04 to 0.01Å) so its interaction with a slice of a crystal will generate a diffraction grating just as it happens in the case of visible light and X-rays. Unlike X-rays, it is possible to focus an electron beam by placing suitable electromagnets along the passage of the beam. However, most crystals are opaque for the electron beam at the thickness of a normal optical thin section (30ì m). Hence, special techniques are employed to prepare suitable samples (~0.1ì m thickness) of mineral specimens for TEM investigations, *e.g.*, *acid jet spray* or *ion milling grinder*. The TEM studies are eminently suitable for exploring intergrowths, crystal defects, diffusion of ions etc.

Design of a TEM

Fig. 10.7 gives a schematic view of a TEM which essentially resembles a microscope tube (Chapter 5 Page YY) though a giant one. The tube or column, as it is popularly called, consists of six components; all these are connected to a series of three or four different vacuum pumps that work in a serial order progressively ensuring a very high degree of vacuum. At the very top, (A in Fig. 10.7) is the electron gun that generates an electron beam. The beam passes through a set of condenser lenses that focuses it on the object placed on the specimen stage (B in Fig. 10.7). The beam is transmitted through the object generating a diffraction pattern which is collected by the objective lenses (C in Fig. 10.7). The extent of the diffraction pattern collected depends upon the width of the aperture placed at the back focal plane of the objective lens. After the diffracted beam passes through the objective, it is focused by the projector

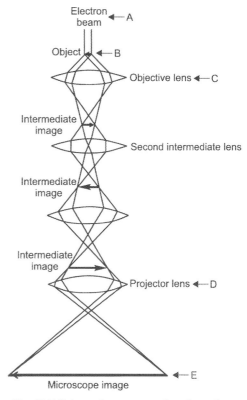

Fig. 10.7 Schematic cross-section through a column of a transmission electron microscope.

lenses (D in Fig. 10.7), equivalent of the Huygens ocular, which focuses image on a special screen (E in Fig. 10.7). The magnification of the image is determined by the projector lenses. The TEM can function on the diffraction mode where diffraction pattern is obtained or, on image mode where the image of the reciprocal lattice of the mineral slice is obtained.

Applications of TEM

TEM has come to be recognized as routine equipment for mineral studies; therefore, its applications in mineral sciences are many. As a typical illustration of TEM applications that is relevant to feldspars (Chapter 15), we may take a study that involves substitution of Sr in the structure of feldspars[4]; it is important to know evolution of microstructures as the mineral is synthesized and annealed to room temperature. Electron diffraction pattern of strontium rich calcic plagioclase is shown in Fig. 10.8. The pattern shows the presence of Carlbad twinning; there are two twin axes (Fig. 10.8A): [100] and [001]. The prominent streaking in Fig. 10.8A and B is due to polysynthetic twining. Dark-

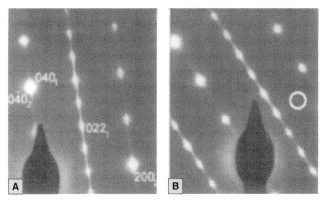

Fig. 10.8 A. Electron Diffraction pattern of a synthetic Sr-rich calcic plagioclase. B. Same sample as that in A, but with a comparatively longer annealing period. (With the kind permission of Mineralogical Magazine; after Trinaudino et al. 2009)

field image of this sample is shown in Fig. 10.9A with antiphase domains shown in Fig. 10.9B. It is also possible to get a reasonable idea of unit cell patterns, linear and planar defects through High Resolution Transmission Microscopy (HRTEM), such as shown in Fig. 10.10 A with the corresponding electron diffraction pattern in Fig. 10.10B of an antigorite sample[5].

---

4. Tribaudino et al. 2009.
5. Cresey et al. 2008.

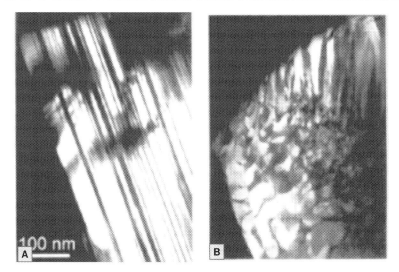

Fig. 10.9 A. Dark-field image of Carlsbad twinning. **B.** Dark field-image of antiphase domains. (With the kind permission of Mineralogical Magazine; after Trinaudino et al. 2009)

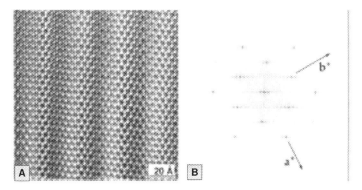

Fig. 10.10 A. An HRTEM images of an antigorite sample. **B.** Electron diffraction pattern of the same sample. (With the kind permission of Mineralogical Magazine; after Cresey et al. 2008)

Scanning Electron Microscope (SEM)

The principle of a *scanning electron microscope* or, SEM, is quite different from that of TEM although both microscopes have electron beam gun. The energy of the electron beam in the case of SEM is much less (~30kV) in comparison to that in the TEM (100-1000kV). Instead of transmission of the beam, as in TEM, the beam hit the specimen along a line in SEM; then the line moves continuously parallel to itself scanning the entire object; hence, the name: scanning electron microscope. When a beam of electrons strikes the surface of a given object, then two phenomena take place. The first reaction is the outcome of collision of high energy electrons with the electrons of the object in the form of ejection of low energy *secondary electrons*. The second reaction is the reflection of the incident electrons after interacting with the atoms of the object; these are called *backscattered electrons* (BSE). Surface topography of the object controls the intensity of secondary and backscattered electrons. Atomic number of the element whose atoms interact with the incident beam also control the intensity of the backscattered beam.

Design of a SEM

Fig. 10.11 illustrates the design of major components of a SEM. There is a source of electrons at A ('electron gun') from which an electron beam is made to pass through a set of scanning coils (S) that are connected to a scan generator (G). The attached detector (D) collects the electrons from the sample, and these signals are amplified to the desired extent by signal amplifiers (H). The entire assembly is shown in Fig. 10.12. An interaction occurs between a cathode ray tube and the signals. This permits us to prepare a high magnification photograph or monitor screen view of the surface of the sample. In a sense it is analogous to the reflecting light microscope. By varying the power of a detector, one can get a secondary electron image or a backscattered image.

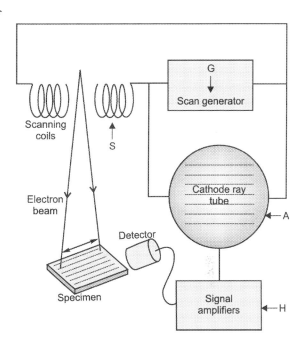

Fig. 10.11 A simplified sketch of the design of a scanning electron microscope.

Applications and Results

It has already been mentioned above that atomic number of an atom in a sample controls intensity of the radiation. It is illustrated in Fig. 10.13 which is back-scattered picture of a garnet porphyroblast from the eclogite of Ladakh in which one can easily mark out different minerals. Since the investigator has already seen the thin section, it is easy to mark which inclusion grain is phengite, calcic amphibole, calcic pyroxene, plagioclase and rutile. Because of titanium, rutile forms the brightest spot and quartz the darkest.

186   Optical Mineralogy

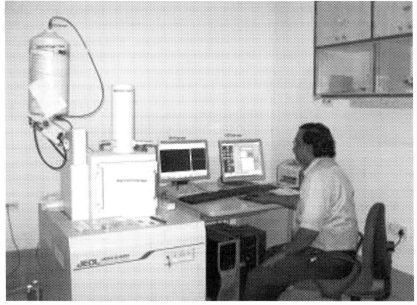

(Courtsey: N.C. Pant)

Fig. 10.12 The scanning electron microscope assembly at Indian Institute of Technology, Kharagpur.

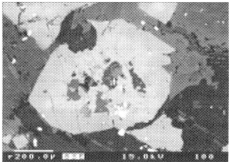

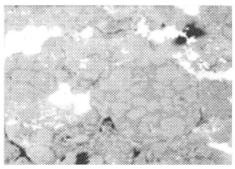

(Courtsey: P. Singh)   (Courtsey: F. Beroud, Laboratoire de Science de la Terre, UCB-Lyon ENS- Lyon)

Fig. 10.13 BSE picture of a garnet grain from an eclogite sample from Ladakh. Contrasting properties of inclusion minerals bring out very clearly inclusion host relationships.

Fig. 10.14 BSE picture of a chondrite. Note the clarity with which internal structure of a chondrule is shown. Bright regions are Fe/Ni metal. www.meteoriteshow.com

These BSE pictures are being increasingly substituted for photomicrographs because of their clarity and high magnification. Fig. 10.14 brings out very vividly the textural relationships in a meteorite sample in another BSE picture. There has been considerable advancement in recent years in SEM techniques and a false colored photograph of a BSE image can also be given. The relationship in inclusions is brought out conspicuously for interpreting the mineral paragenesis in a BSE image. Secondary electron images procured from a SEM can give 3-D perspective very clearly. For example, in Fig. 10.15[6] a SE image was taken for a study on the weathering mechanism of feldspar, an important subject for sedimentologists, geomorphologists, and of course for mineralogists. It shows the surface of the grain that has developed a clay crystal (C) along with amorphous weathering product (WP).

Another common application of SEM that is being routinely carried out is to first take a BSE by using x-ray emission of

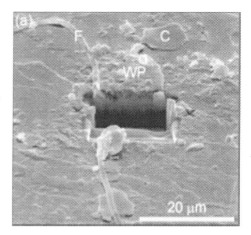

Fig. 10.15 SE image of a feldspar grain tilted at 50 away from the viewer. The feldspar surface has shallow steps and is partly overlain by structureless weathering products (WP). (With the kind permission of Mineralogical Magazine; after Lee et al. 2008)

elements and map the surface of the grains for the concentration of a particular element. For example, in Fig. 10.16 Mn concentration map of a garnet grain is given which clearly shows that there is a pronounced and continuous zoning with reference to manganese concentration in this grains. This prior knowledge helps us

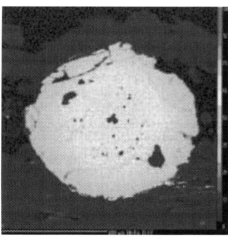

(Courtsey: N.C. Pant)

Fig. 10.16 Mn concentration map of a garnet grain.

in the planning of obtaining quantitative concentration data by EPMA.

Most present day SEM's are equipped to act as *cathodoluminscence* spectrometer as well; though cathodoluminscence spectrometer can also be acquired as an accessory to a polarizing microscope also. Cathodoluminescence (CL) is the emission

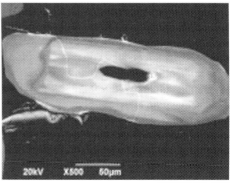

(Courtesy: N.C. Pant)

Fig. 10.17 A CL picture illustrating zoning in a zircon grain.

---

6. Lee et al. 2008.

Atomic Force Microscopy (AFM)

of visible light from the specimen due to its interaction with the primary electron beam. It arises due to point defects in the lattice, and, the presence of large elements, (*e.g.*, lanthanides) which might manifest effects similar to point defects. Regardless of its genesis, the CL technique offers us an insight, from example, zoning (Fig. 10.17), that can indicate areas of special interest within the specimen.

Atomic Force Microscopy (AFM)

In early years of 1980's, a *scanning tunneling microscope* was designed by Gerd Binning and Heinrich Rohrer who modified it to what is now known as AFM. Just as a scanning electron microscope (SEM) images the surface texture of a sample the *atomic force microscopy* (AFM) has a microprobe that scans the surface of the sample. The major difference between the SEM and AFM is that latter can supply information not only about the surface texture but also about any other physical and chemical property, such as magnetic domain variation. An AFM image is also a true 3-D image whereas the image by SEM is essentially a two dimensional projection. Moreover, it is easy to maintain an AFM as compared to a SEM because the AFM can work in an ambient air or air-liquid environment. Additionally, no gold or carbon coating is required for viewing samples under AFM. However, the AFM suffers from the drawback of sample size. While SEM may be used for millimeter scale samples with millimeter level depth, the AFM is suitable for micron or nannoscale coverage. Another point that must remembered by the AFM users is that image reconstruction is slow in this case but in SEM it is accomplished in real time.

In Fig 10.18, illustrates the principle of this technique. You would notice that there is a *cantilever beam* (P) at one end of which a small probe (diameter < 100 microns) is attached (S in Fig. 10.18). The beam (P) is made up of silicon or silicon nitride with a radius of curvature on the nanometer scale. The probe scans the sample over a designated area along X-Y. The probe is sensitive to *interatomic forces*. These forces consist of a number of such as components *mechanical contact force*, *van der Waal* force, *capillary forces*, *bonding forces*, *electrostatic forces* etc. Therefore, as the scanning progresses, the atoms of the sample and those of the probe interact with each other causing the cantilever to deflect in accordance with the surface topography or some other chosen property of the sample. Considering the variation in topography, the interatomic forces would recreate the variation and supply the data to a computer where an application package would show graphically this difference (Fig. 10.19, also see the title page). The deflection of the cantilever is monitored through a laser spot on the top of the lever deflected towards a set of photodiodes. Normally, the area covered is quite small, *e.g.*, 100 microns but it could be as small as 100nm. The sample mount is commonly a *piezoelectric tube* (C) that is capable of moving in Z direction. The Z movement is guided by a 'feedback' system that maintains a constant force between the probe and the sample. In some designs, the sample mount consists of three piezoelectric crystals that take care of X, Y, and Z directions of scanning. In still another variation of the

New Frontiers in Microscopy

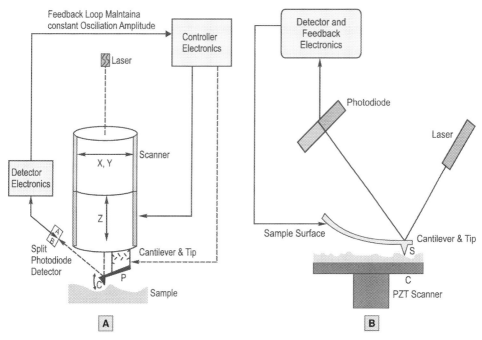

Fig. 10.18 A sketch of AFM. **A.** AFM instrument configuration; **B.** Details of AFM sensor.

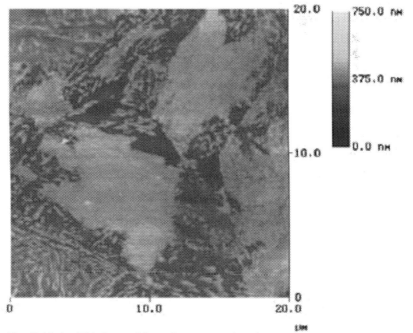

Fig. 10.19 An AFM picture of the surface texture of steel.

sample mount-probe, the probe is mounted on the vertical *piezoscanner* and the sample has only X and Y *piezocrystals*.

Applications of AFM

The AFM can be used in several modes. For example, one uses it in contact mode. In this a constant force is maintained between the probe and the sample. This is a common technique for topographic mapping where a height difference of even 1Å is easily measured. In the intermittent mode the probe cantilever is made to oscillate near its resonance frequency. As the probe is brought close to the surface, the vibrational amplitude is affected by the topography and other physical properties of the sample. This information is transmitted to the computer for preparation of an accurate graph of the property (Fig. 10.19). Similarly, the AFM can be set up in lateral mode where the probe measures the lateral deflection of the cantilever. It is based on the concept that frictional force between the probe and the sample surface would vary according to the surface irregularities. It is also possible to monitor the phase lag between the signals of the drive and cantilever oscillations. It is known as the phase detection microscopy and is best suited to study mechanical properties of different parts of sample surface.

Summary

In several rocks mineral grains carry minute inclusions that are actually cavities filled with solid and fluid phases. Investigations on these inclusions involve a polarizing microscope fitted with an accessory stage called heating and freezing stage or microthermometric stage. Temperatures in the range of -196°C to 600°C may be generated within the sample holder of the stage and can be effectively controlled. The changes in temperature causes phase changes that allow us to interpret the nature of the fluids trapped inside the cavity. Statistical analysis of various measurable parameters of an image of mineral grains is the subject of Image Analysis. Minerals are identified on the basis of their reflectivity and assigned names accordingly. Different grains of each mineral are measured for their assigned parameters and output is saved on the hard disk for further manipulations.

Increasing use is being made of Transmission Electron Microscopy and Scanning Electron Microscopy in mineral studies. The former furnishes information on internal structure of the mineral. It requires special sample preparation where the thickness is reduced to 10Å. TEM works in both diffraction and picture modes. SEM scans the surface of the mineral to provide vital information about the surface features of the grains. There is a provision in SEM to switch to cathodoluminiscence mode. Atomic Force Microscopy also provides the information on the surface features of rocks and minerals but on much smaller scale giving extreme details about a small part.

Important Terms

acid jet spray 182
atomic force microscopy 188
backscattered electrons 185
bonding forces 188
camera lucida 178
capillary forces 188
cathodoluminscence 188
electrostatic forces 188
fluid inclusions 176
heating-freezing stage 177
image analysis 178
interatomic forces 188
ion milling grinder 182
Ion Microprobe 176

mechanical contact force 188
mechanical stage 178
melt inclusions 176
microthermometric stage 177
ocular graphs 178
piezoelectric tube 188
piezoscanner 190
piezocrystals 190
point counter 178
scanning electron microscope 184
secondary electrons 185
scanning tunneling microscope 188
transmission electron microscope 182
van der Waal force 188

Questions

1. Define the term fluid inclusion.
2. Describe the common type of fluid inclusions.
3. What is standard method of studying fluid inclusions?
4. What is the purpose of image analysis?
5. Enumerate some of the applications of the image analysis system.
6. What is modal analysis and how would you carry out with the help of the image analysis system?
7. How would you carry out grain size analysis with the help of the image analysis system?
8. What is meant by Transmission Electron Microscopy?
9. What are the applications of TEM in mineralogical investigations?
10. Describe different procedures of sample preparation for TEM studies.
11. Briefly describe important components of TEM with the help of a simple sketch.
12. What are major differences in Transmission Electron Microscopy and Scanning Electron Microscopy?
13. What is meant by secondary electrons in the context of SEM?
14. What is meant by back-scattered electrons in the context of SEM?
15. What is a BSE pictures? What are its advantages over an optical photomicrograph?
16. What is the basic principle of atomic force microscopy?

17. Compare the advantages of AFM and SEM.

18. Briefly describe important components of AFM.

19. What are important applications of AFM in mineral investigation?

Suggestion for Further Reading

Bakker, R.J., 1999, Adoption of Bowers & Helgeson (1983) equation of state to isochore and fugacity coefficient calculation in the H_2O-CO_2-CH_4-N_2-NaCl fluid system, Chemical Geology, 154, 225-236.

Bakker, R.J. and Diamond, L.W., 1999, Determination of the composition and molar volume of H_2O-CO_2 fluid inclusions by microthermometry, Geochimica et Cosmochimica Acta.

Bakker, R.J. and Mamtani, M.A., 1999, Fluid inclusions as post-peak metamorphic process indicators in the Southern Aravalli Mountain Belt (India), Contributions to Mineralogy and Petrology.

Cresey, G, Cresey, B.A. and Wicks, F.J., 2008. Polyhedral serpentine: a spherical analogue of polygonal serpentine, Min. Mag., 72, 1229-1242.

Giessibl, F., Advances in Atomic Force Microscopy, Reviews of Modern Physics, 75 (3), 949-983, 2003.

Lee, M.R., Brown, D.J., Hodson, M.E., Mackienzie, M. and Smith, C.L., 2008, Weathering microenvironments on feldspar surface: implications for understanding fluid-mineral reactions in soils. 72. 1319-1328.

McLaren, A.C., 1991, Transmission Electron Microscopy of minerals and rocks, Cambridge University Press.

Morris, V.J., Kirby, A.R. and Gunning, A.P., 1999, Atomic Force Microscopy for Biologists, Imperial College Press.

Tribaudino, M. Zhang, M. Salje, E.K.H., 2009, Cation ordering and phase transitions in feldspars along the join $CaAl_2Si_2O_8$-$SrAl_2O_8$: a TEM, IR and XRD investigations, Min. Mag., 73, 119-130.

PART II

Systematic Description of Common Rock Forming Minerals

11 Nesosilicates

- Olivine
- Monticellite
- Garnet
- Vesuvianite
- Zircon
- Andalusite
- Kyanite
- Sillimanite
- Mullite
- Staurolite
- Sphene
- Topaz
- Chloritoid

Technical Text Box 11.1

The nesosilicates (orthosilicates of many authors) are characterized by the presence of $[SiO_4]^{4-}$ tetrahedra that do not share any of their four oxygen atoms with any other $[SiO_4]^{4-}$ present in the unit cell. Other metal ions of a given nesosilicate may form octahedra, (*e.g.*, Mg, Fe as in olivine) and/or cube, (*e.g.*, as in garnets), or even five fold coordination, (*e.g.*, as in andalusite) or another group of tetrahedra, (*e.g.*, as in sillimanite). Zirconium ion in zircon has a complex cubic coordination.

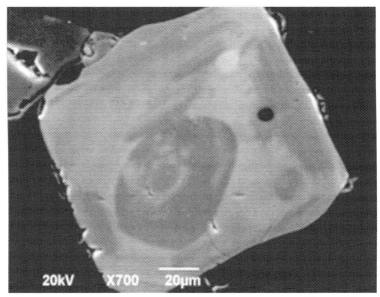

(Courtesy: N.C.Pant)

A cathodluminescence image of a zoned zircon.

Olivine

Nature of the Phase

Essentially a binary solid solution; end members are Forsterite (Mg_2SiO_4) and Fayalite (Fe_2SiO_4).

Crystallography

Orthorhombic; 2/m2/m2/m; Pbnm.

Z = 4	Forsterite(fo)	Fayalite(fa)
á =	1.635	1.827
â =	1.651	1.869
ã =	1.670	1.879

The refractive indices vary linearly with composition as expressed as $Fa_\%$ (Fig. 11.1). The ideal graph may be complicated by the presence of zoning.

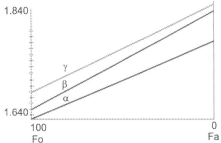

Fig. 11.1 Variation of refractive indices with $Fa_\%$ in olivine.

Plane Polarized Light

Form and Relief: Anhedral to subhedral rounded sub rounded grains (Fig. 11.2A); as phenocrysts it occurs as eight or six sided grains (Fig. 11.2B). High relief; increasing with increase in iron content. Internal fracturing is common; fractures often filled with serpentine minerals. May contain inclusions of opaques.

In many rocks, such as, komatiite, olivine may occur as elongated, skeltal (Fig. 11.2C), and dendritic in nature, giving rise to spinifex texture.

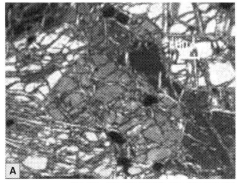

(Courtesy: Department of Geology Collection, DU)

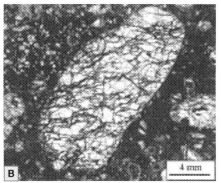

(Courtesy: S.C. Patel) (Courtesy: D. Walker)

Fig. 11.2 Forms of olivine. **A.** Equigranular rounded olivine grains; **B.** Nearly euhedral olivine grain; **C.** Skeltal habit of olivine.

Color and Pleochroism: Mg olivine is colorless to pale brown. With increasing Fe content, the color tends to acquire darker shades of yellow. High Fe content may induce pleochroism in an otherwise colorless grains. Pleochroic Scheme: á = ã = pale to lemon yellow; â = orange.

Cleavage: Commonly not observed. In large grains poor to imperfect on {010} and {100}. The quality of cleavage improves with increasing Fe content, particularly {010}. Irregular fracture pattern.

Under Crossed Polars

Anisotropism: Anisotropic.

Birefringence: Moderate to strong. Best seen in the basal section.

ã: 0.035(fo)   0.052(fa).

Polarization Colors: Second order colors.

Extinction: In most cases, it is not possible to get a good straight edge or a cleavage set to measure extinction angle. The internal structure of olivine is characterized by the presence of slip surfaces. Hence, translational gliding gives rise to lamellar structure; inhomogenous gliding creates kink bands. Such features give rise to undulose extinction.

Orientation: á = y, â = z, ã = x for the entire range of composition (Fig. 11.3).

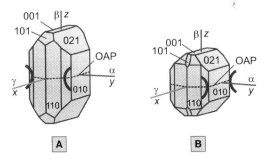

Fig. 11.3 A. Optic Orientation of Forsterite; **B.** Optic Orientation of Fayalite.

OAP: (001) for the entire range of composition.

Indicatrix: Forsterite(fo)   Fayalite(fa)
              Biaxial (+)   Biaxial (−)

For fo-rich end (100), and for fa-rich end (010) sections are suitable for obtaining acute bisectrix figures. Single isogyre is easy to obtain.

$2V_a$: Forsterite(fo)   Fayalite(fa)
           82°   134°

$2V_a$ is very large. Increases with the addition of Fe content. When the composition is Fa_{13} $2V_a$ becomes 90°, the mineral becomes biaxial negative. There is a straight line relationship between $2V_a$ and fa%; hence, it provides a quick and reliable method for Fa estimation (Fig. 11.4).

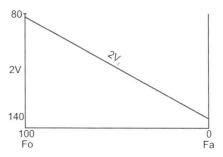

Fig. 11.4 Variation of $2V_a$ of olivine with its $Fa_\%$.

Sign of Elongation: Often olivine occurs as rounded or sub-rounded grains; when elongated prismatic sections are available, these could be length fast or length slow because elongation is parallel to z direction which coincides with â.

Dispersion: Weak; r > v over á.

Twinning: {011}, {012}, {031}.

Zoning: Seen in phenocrysts easily; Mg-rich core with rim of Fe-rich. Easily distinguished by the differences in birefringence of different zones. No hasty conclusions about Fa% should be made, since (010) and (100) sections would show contrasting birefringence behaviour. See Figs. 11.3A and B.

Alteration

Very common. The product is a complex mixture of clay-size minerals. There are three main groups of alteration products. The first is *serpentine* that is common in many igneous rocks. It begins along fractures and proceeds inwards to, at times, cover the entire grain (serpentine pseudomorph after olivine). The product minerals are fibrous, commonly of chrysotile, antigorite, talc. Carbonates, particularly, magnesite, may also be present along with magnetite. The second group of alteration product is called *Iddingsite*. It appears as fine-grained mixture of reddish brown minerals with high refractive index and high birefringence. *Chlorophaeite* is another alteration product that is isotropic and of orange to green in color.

Distinguishing Characters

Pyroxenes and epidotes bear similarities with olivine. But both these minerals have very good cleavages. A (010) section of pyroxene and of epidote shows inclined extinction. Pyroxenes have lower birefringence. The fracture pattern and serpentine alteration would also reveal the presence of olivine.

Mineral Association

Commonly in igneous rocks, but also in marbles. As an igneous mineral it occurs with spinel, ortho-, and clinopyroxenes and calc-plagioclase. Opaques like, magnetite and chromite are generally associated. Dunite essentially composed of olivine; Fe-end member variety, i.e., fayalite is reported from many granites. In marbles, forseterite rich olivine is associated with serpentine, diopside etc.

Monticellite

Nature of the Phase

A binary solid solution between $CaMgSiO_4$ and $CaFeSiO_4$; in nature Mg rich phase is common and is always close to stoichiometric composition.

Crystallography

Orthorhombic; 2/m2/m2/m; Pbnm.

> Z = 4
> á = 1.638 – 1.654
> â = 1.646 – 1.664
> ã = 1.653 – 1.674

Plane Polarized Light

Form and Relief: Anhedral small rounded grains. Relief moderate.

Color and Pleochroism: Colorless and non-pleochroic.

Cleavage: Commonly not seen; {010} poor cleavage.

Under Crossed Polars

Anisotropism: Anisotropic.

Birefringence: Moderate. Best seen in the basal section.

ä: 0.012 – 0.020.

Polarization Colors: Up to middle first order colors.

Extinction: Straight extinction.

Orientation: á = y, â = z, ã = x (Fig. 11.5).

OAP: (001).

Indicatrix: Biaxial (–). Interference figure difficult to obtain; few isochromes.

$2V_a$: ≤ 90°.

Dispersion: Distinct; r > v.

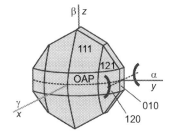

Fig. 11.5 Optic orientation of Monticellite.

200 Optical Mineralogy

Twinning: {031}, cyclic.

Zoning: Commonly not seen.

Alteration

Not common. At times may alter to serpentine or augite; may partly change to merwinite or larnite.

Distinguishing Characters

Can be confused with diopside but monticellite has no cleavage. Lower birefringence than forsterite; mineral association with carbonates is also characteristic.

Mineral Association

Contact aureole where carbonate rocks are common.

Garnet

Nature of the Phase

A multicomponent complex solid solution. General formula $A_3B_2Si_3O_{12}$, where A is a divalent metal ion, and B is a trivalent metal ion. Table 11.1 list common end-members of this solid solution. Empirically, on the basis of common occurrence in nature, two subgroups of this group are suggested and listed. In addition, garnet has been found to contain a large number of trace elements including rare earths.

Table 11.1 Common garnet minerals and their chemical composition

Pyralspite Group		Ugrandite Garnets	
Pyrope	$Mg_3Al_2(SiO_4)_3$	Uvarovite	$Ca_3Cr_2(SiO_4)_3$
Almendine	$Fe_3Al_2(SiO_4)_3$	Grossular	$Ca_3Al_2(SiO_4)_3$
Spessartine	$Mn_3Al_2(SiO_4)_3$	Andradite	$Ca_3Fe_2(SiO_4)_3$

Crystallography

Orthorhombic; $4/m\,\overline{3}\,2/m$; Ia3d.

$$Z = 8$$

Pyralspite Group	n	Ugrandite Garnets	n
Pyrope	1.714	Uvarovite	1.865
Almendine	1.830	Grossular	1.734
Spessartine	1.800	Andradite	1.887

Nesosilicates

Plane Polarized Light

Form and Relief: Commonly euhedral, may be fine-grained, but generally medium to coarse-grained. Common forms are dodecahedral {011} and trapezohedral {112}. Relief strong (Fig. 11.6). In several instances irregular rounded grains or skeletal.

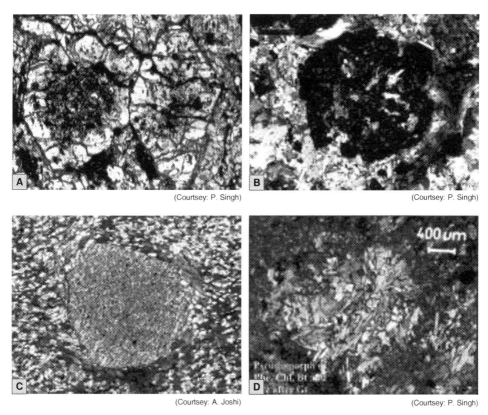

Fig. 11.6 Habits of garnet grains. **A.** Garnet grains in the matrix of amphiboles. Plane polarized light; **B.** A euhedral garnet under crossed polars; **C.** A large garnet grain (porphyroblast) within a matrix of a metapelite. Note large number of quartz grains that occur as inclusions within the garnet grain. All the inclusions are parallel to each other and represent a fabric in the rock before the growth of garnet. Post-garnet growth fabric wraps around the porphyroblast; **D.** Chlorite occurring as pseudomorph after garnet.

Color and Pleochroism: Generally non-pleochroic, though strained varieties may show some pleochroism

Cleavage: Commonly not observed. Dodecahedral parting {011} rare, conchoidal or subconchoidal fractures very common.

Under Crossed Polars

Anisotropism: Isotropic. However, strained grains, because of internal fracturing and annealing may show anisotropism. It is particularly true for garnets rich in Ca and Mn.

Zoning: Fairly commonly zoned on the level of concentration of bivalent metal ions. It also is manifested, at times, in color zoning (Fig. 11.6A). *See* also Fig. 10.16. Garnet in garnet giving an impression of zoning are also seen in many metamorphic rocks.

Inclusions: Very common. May show patterns, like, linear, spiral, crenulations etc. (Figs. 11.6A, B and C, Fig. 10.13). Quartz is the most common mineral as inclusions, followed by mica, plagioclases and other matrix minerals

Alteration

Very common. It may begin along numerous fractures that are commonly present, or it may start from the circumference of rounded grains. There may be multi-layered rings, called coronas, around garnet grains. Main alteration product is chlorite (Fig. 11.6 D), but amorphous oxides of iron, manganese also found. In specific examples, epidote or even hornblende may occur.

Distinguishing Characters

High relief, euhedral grain shape and isotropic characters allow garnet to be identified easily.

Mineral Association

Many index minerals of Barrovian zones are common associates, like chlorite, biotite, staurolite, kyanite, sillimanite etc. Cordierite, epidote are common along with hornblende, plagioclase, sphene etc. Garnet is an essential mineral of garnet peridotite, an upper mantle rock, where it occurs with olivine and orthopyroxenes. Common as a detrital mineral in sandstones.

Vesuvianite (Idocrase)

Nature of the Phase

The composition is fairly constant with a composition of $Ca_{19}(Al,Fe)_{10}(Mg,Fe)_3[Si_2O_7]_4$ $[SiO_4]_{10}(O,OH,F)_{10}$.

Crystallography

Tetragonal, 4/m2/m2/m, P4/nnc.

Nesosilicates

$$Z = 2$$
$$\mathring{a} = 1.700 - 1.746$$
$$\grave{u} = 1.703 - 1.752$$

Plane Polarized Light

Form and Relief: It occurs as prismatic crystals, but in general, crystals are subhedral; only a few faces are present. Relief is high; relief increases with increasing amount of titanium and iron.

Color and Pleochroism: Colorless, pale yellow or pale brown. Colored varieties may show weak pleochroism from brownish yellow to weak yellow.

Cleavage: {110} poor, {100} and {001} very poor.

Under Crossed Polars

Anisotropism: Anisotropic, but basal section is isotropic.

Birefringence: Low; decreases with increasing (OH) content; may become isotropic or optically positive for a particular wavelength.

ä: $0.001 - 0.009$.

Polarization Colors: First order grey; often anomalous blue.

Extinction: Straight extinction.

Indicatrix: Uniaxial (−).

Sign of Elongation: Prismatic sections are length fast (negative elongation).

Dispersion: Strong.

Twinning: Rare.

Zoning: Not seen.

Alteration

Not seen.

Distinguishing Characters

The tetragonal form, high relief and low birefringence are characteristics of vesuvianite. It resembles zoisite but usually occurs as large mineral grains; may resemble grossularite.

Mineral Association

An important calc-silicate mineral, common in marble. Ca-garnets, wollastonite are common associate; may also occur in igneous rocks such as syenites and pegmatites.

Zircon

Nature of the Phase

Treated as a pure phase with stoichiometric composition of $ZrSiO_4$. Many natural grains invariably contain a small amount of uranium and other radioactive elements.

Crystallography

Tetragonal, 4/m2/m2/m; I4$_1$/amd.

$$Z = 4$$
$$\grave{u} = 1.922 - 1.960$$
$$\mathring{a} = 1.961 - 2.015$$

Plane Polarized Light

Form and Relief: Zircon exhibits different habits in different kinds of rocks. In igneous rocks zircon grains are commonly euhedral; prismatic sections show outline of prisms terminated at both ends by pyramids in accordance with its high symmetry (Fig. 11.7A). In sedimentary rocks, the mineral occurs as rounded to subrounded grains. Metamorphic zircons are invariably zoned; the zoning may be due to euhedral overgrowth over rounded detrital zircons, or zircon formed by more than one phase of metamorphism (Fig. 11.7B; also Fig. 10.17). Zoning may also arise due to composition variations of minor elements. Another common habit is as inclusions in other minerals, commonly in cordierite or biotite (Fig. 11.7C).

As mentioned above, almost as a rule radioactive elements like U and Th substitute for Zr in the structure of zircon. The decay of radio-isotopes progressively destroys the mineral structure; additionally, it also burns the neighbourhood region of the grain (Fig. 11.7C). Hence, in most cases, true relief is masked by these changes. Otherwise the grains form prominent relief.

Color and Pleochroism: Colorless and non-pleochroic.

Cleavage: Commonly not observed.

Under Crossed Polars

Anisotropism: Anisotropic; basal sections are isotropic.

Birefringence: Strong birefringence; seen best in elongated prismatic section

$\mathring{a}$: $0.042 - 0.065$.

Polarization Colors: Bright third or fourth order colors.

Extinction: Parallel extinction.

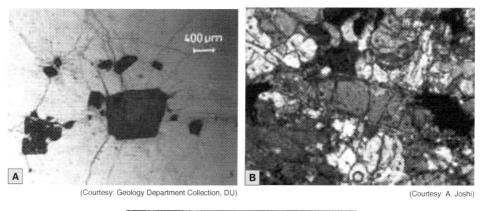
(Courtesy: Geology Department Collection, DU)   (Courtesy: A. Joshi)

(Courtesy: Geology Department Collection, DU)

Fig. 11.7 Habits of zircon. **A.** Euhedral zircon grains; prismatic and basal sections are seen; radial fractures around a large grain are due radiations emanating from uranium included in the zircon structures; **B.** Zoned zircon; **C.** Inclusions of zircon in a biotite laths; a pleochroic halo is easily seen.

Indicatrix: Uniaxial (+).

Sign of Elongation: c-axis shows positive elongation, *i.e.*, prismatic section is length slow.

Dispersion: Weak; $r > v$ over á.

Twinning: Not common.

Zoning: Very common. It may often be due to overgrowth. Zircon survives as a detrital mineral in sediments. These sediments are metamorphosed, even partially melted during a later geologic event. Zircon not only survives, but the detrital grains act as seeds for further growth (*see* Fig. 10.17). Another common reason for zoning is the decay of incorporated uranium in its structure. The emitted radiation causes destruction radially from the decay site causing a zonal pattern.

Alteration

Due to radioactive decay as mentioned above. Such grains are called metamict zircons. Otherwise always fresh.

Distinguishing Characters

Tiny elongated and euhedral crystals with high relief and high birefringence.

Mineral Association

Common mineral in granites, pegmatites and other acidic igneous rocks. Common as inclusions in biotite forming pleochroic halos. Common inclusion in garnets of metamorphic rocks.

Andalusite

Nature of the Phase

The mineral is relatively pure with stoichiometric composition of Al_2SiO_5, and the only other ions present in noteworthy amounts are ferric iron and manganese.

Crystallography

Orthorhombic, 2/m2/m2/m, Pnnm.

$$Z = 4$$
$$á = 1.633 - 1.642$$
$$â = 1.639 - 1.644$$
$$ã = 1.644 - 1.650$$

Plane Polarized Light

Form and Relief: Commonly occurs as euhedral elongate prisms (Fig. 11.8A); prisms may have a square cross-section, with a moderate relief. The square cross-sections, many a time, reveal a cross like arrangement of graphite grains (Fig. 11.8B). This variety is known as *chiastolite*.

Color and Pleochroism: Usually colorless. However, some occurrences of andalusite are strongly colored in pale or pink. Such sections would exhibit pleochroism with á = pink, yellow â = colorless or pale yellow and ã = colorless, greenish yellow.

Cleavage: {110} good, appearing as traces parallel to the prism edge in the prismatic sections but intersecting at right angles in a basal section $(110) \wedge (1\overline{1}0) = 89°$ {100} poor.

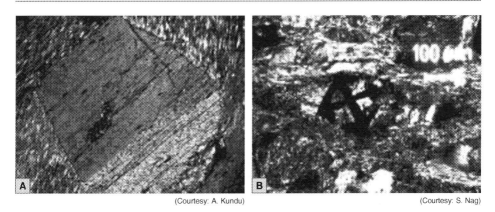

(Courtesy: A. Kundu)   (Courtesy: S. Nag)

Fig. 11.8 Photomicrographs of andalusite. **A.** The grain is twinned. Under cross polars; **B.** A schist which suffered contact metamorphism prior to an episode of regional metamorphism. Note chiastolite cross section of andalusite.

Under Crossed Polars

Anisotropism: Anisotropic.

Birefringence: Low; best seen in (010) section.

ã: 0.009 – 0.012.

Polarization Colors: First order grey.

Extinction: Straight on prism edge or on {110} cleavages. Basal sections give symmetrical extinction.

Orientation: á = z, â = y, ã = x (Fig. 11.9). However, when $Mn^{3+}+Fe^{3+}$ is > 6.5 mol% orientation changes to á = x, â = y, ã = z.

OAP: (010) for all composition ranges.

Indicatrix: Biaxial (–),

$2V_á$: 73° – 82°.

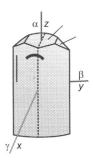

Fig. 11.9 Optic orientation of andalusite.

Basal sections give acute bisectrix optic figure. Cleavage fragments yield off-centred or flash figures.

Sign of Elongation: Negative, length fast.

Dispersion: Weak, $v > r$.

Twinning: Rare on {101} (Fig. 11.8A).

Zoning: Uncommon, Fe^{3+} distribution is usually the cause.

Alteration

Andalusite to sillimanite and sericite common.

Distinguishing Characters

Almost square cross-section, moderate relief and low birefringence are characteristics of this mineral. Black cross (chiastolite variety) seen in cross-section. Parallel extinction in prismatic section distinguishes it from kyanite.

Mineral Association

It is a common contact metamorphic mineral along with biotite and cordierite. It occurs with garnet or staurolite in several regional metamorphic terrains.

Kyanite

Nature of the Phase

Treated as a pure phase with stoichiometric composition of Al_2SiO_5

Crystallography

Triclinic, $\overline{1}$, P $\overline{1}$.

$$
\begin{aligned}
Z &= 4 \\
á &= 1.710 - 1.718 \\
â &= 1.719 - 1.724 \\
ã &= 1.724 - 1.734
\end{aligned}
$$

Plane Polarized Light

Form and Relief: Bladed habit is common (Fig. 11.10A) as subhedral triclinic pinacoids broad in direction and thin in a direction normal to it (Figs. 11.10B and C). Radiating aggregates of blades also common (Fig. 11.10D). High relief (Fig. 3.2B, page YY).

Color and Pleochroism: Usually colorless in thin sections but may be pale blue. Pleiochroism is weak, but seen in thick sections with á = colorless, â = ã = blue.

Cleavage: {110} perfect; {010} good, {001} partings.

Under Crossed Polars

Anisotropism: Anisotropic.

Birefringence: Low.

ã: 0.012 – 0.016.

Nesosilicates

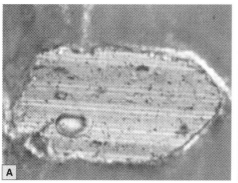
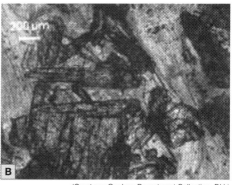

(Courtesy: S.J. Patel)　　　　　　　　　　(Courtesy: Geology Department Collection, DU)

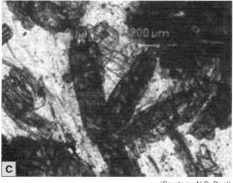

(Courtesy: N.C. Pant)

Fig. 11.10 Photomicrographs of kyanite. **A.** A large kyanite phenocryst from an eclogite sample. Plane polarized light; note polysynthetic twinning; **B.** Kyanite pophyroblast in a metapelite. Crossed polars; **C.** Radiating kyanite blades in a metapelite. Crossed polars.

Polarization Colors: First order grey (similar to quartz).

Extinction: Oblique on cleavage and prism edge; $\tilde{a} \wedge$ prism edge = 30° (Fig. 11.11).

Orientation: $\tilde{a} \wedge z$ on (100) = 27° – 32°, on (010) = 5° – 8°; $\acute{a} \wedge x$ on (001) = 0° – 3° (Fig. 11.11)

OAP: Nearly normal to {100} but inclined by over 30° to both *b* and *c*.

Indicatrix: Biaxial (–).

$2V_{\tilde{a}}$: 73° – 86°. Best figure is seen parallel to major (100) cleavage. Though melatopes may be out of view.

Sign of Elongation: Length slow; positive elongation.

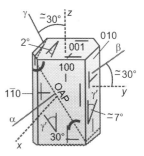

Fig. 11.11 Optic orientation of kyanite.

Dispersion: $r > v$ weak.

Twinning: Lamellar on (100), twin axis perpendicular to (100) or parallel to y or z; multiple on {001}.

Zoning: Commonly not seen. In some cases may be intergrown with staurolite.

Alteration

Kyanite often occurs within large knots of micaceus minerals; it also inverts to sillimanite with increasing temperature.

Distinguishing Characters

The higher birefringence and excellent cleavage and high relief help to distinguish kyanite from andalusite and other index minerals.

Mineral Association

Common regional metamorphic mineral, normally indicative of middle to upper amphibolite facies rocks that are alumina rich. Under extreme conditions also occur in other facies. Quartz, staurolite, and sillimanite are common associated minerals. Common detrital mineral.

Sillimanite

Nature of the Phase

The composition is fairly constant and is close to pure Al_2SiO_5.

Crystallography

Orthorhombic, 2/m2/m2/m, Pbnm.

$$Z = 4$$
$$á = 1.653 - 1.661$$
$$â = 1.657 - 1.662$$
$$ã = 1.672 - 1.683$$

Plane Polarized Light

Form and Relief: It occurs as elongate prisms in two habits: either as small fibrous crystals found in regionally metamorphosed schists and gneisses (*fibrolite*) (Fig. 11.12A), or as small prismatic crystals (Fig. 11.12B). These elongated acicular grains are commonly marked by cross fractures. Relief is moderately high.

Nesosilicates

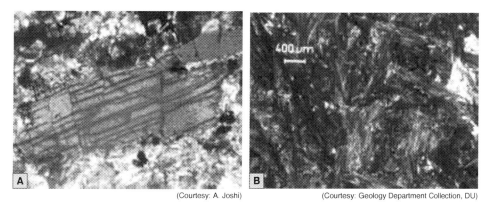

(Courtesy: A. Joshi)   (Courtesy: Geology Department Collection, DU)

Fig. 11.12 Photomicrographs of habits of sillimanite. **A.** Prismatic habit. Note cleavages and cross fractures. Crossed polars; **B.** Fibrolite habit. Crossed Polars.

Color and Pleochroism: Colorless with no pleochroism.

Cleavage: {010} perfect, a basal section of sillimanite, which is diamond shaped, has cleavage parallel to the long axis. Cross fractures may mask the appearance of the cleavage in prismatic laths.

Under Crossed Polars

Anisotropism: Anisotropic.

Birefringence: Moderate. Best seen (010) sections.

ã: 0.018 – 0.022.

Polarization Colors: Bright colors, high first order to low second order.

Extinction: Straight on single cleavage trace.

Orientation: á = x, â = y, ã = z (Fig. 11.13).

OAP: (010).

Indicatrix: Biaxial (+); most fibrous variety samples are too narrow to show figure, and cleavage fragments yield flash figure.

$2V_a$: 21° – 30°.

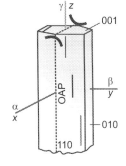

Fig. 11.13 Optic orientation of sillimanite.

Sign of Elongation: Length slow, positive elongation.

Dispersion: Strong, $r > v$ of optic axis.

Twinning: Rare.

212 — Optical Mineralogy

Zoning: Not seen.

Alteration

Rare as sericite.

Distinguishing Characters

The positive elongation and higher birefringence distinguish sillimanite from andalusite, while kyanite has higher refractive indices and a greater 2V.

Mineral Association

A regional and contact metamorphic mineral representing a high grade. Minerals like biotite, muscovite, K-feldspar, quartz, plagioclase, garnet, staurolite are common associates. Rarely with andalusite and kyanite. In granite, it is reported as entrapment of undigested country rock.

Mullite

Nature of the Phase

Mullite is Al silicate with a composition that ranges from $3Al_2O_3.2SiO_2$ (á mullite) to $2Al_2O_3. SiO_2$ (â mullite), show some deviation with 60 mol% Al_2O_3.

Crystallography

Orthorhombic, 2/m2/m2/m, Pbam.

$$Z = 2$$
$$á = 1.630 - 1.670$$
$$â = 1.636 - 1.675$$
$$ã = 1.640 - 1.690$$

Plane Polarized Light

Form and Relief: Commonly occurs as euhedral elongate prisms, prisms may have square cross-sections, with a moderate relief. It may also occur like fibrolite, – an aggregate of fibrous needles.

Color and Pleochroism: Usually colorless or pinkish, when colored show weak pleochroism with á = â = colorless and ã = pinkish.

Cleavage: {010} distinct.

Under Crossed Polars

Anisotropism: Anisotropic.

Birefringence: Low, best seen on (010).

ä: 0.010 – 0.029.

Polarization Colors: First order grey to yellow.

Extinction: Parallel extinction on prism faces.

Orientation: á = x, â = y, ã = z (Fig. 11.14).

OAP: (010).

Indicatrix: Biaxial (+).

2V$_a$: 45° – 61°.

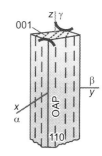

Fig. 11.14 Optic orientation of mullite.

Sign of Elongation: Positive; Cleavage fragments and acicular grains are length-slow.

Dispersion: r > v.

Twinning: Rare.

Zoning: None.

Alteration

Not known.

Distinguishing Characters

Mullite is similar to sillimanite but can be differentiated by cell dimensions and 2V value.

Mineral Association

Not a common mineral in natural rocks though synthetic mullite has extensive use in industry. When found in nature, it occurs in pelitic xenolith in basic igneous rocks.

Staurolite

Nature of the Phase

(Fe^{2+}, Mg, Zn)$_2$(Al, Fe^{3+}, Ti)$_9$O$_6$[(Si, Al)O$_4$]$_4$(O, OH)$_2$. This is a working chemical formula for staurolite because a number of elements such as Mn, Cr, V, Zn etc. often substitute in minor amounts resulting in characteristic changes in physical and optical properties. The following properties belong to samples that are closest to the above-mentioned formula.

Crystallography

Monoclinic (Pseudo-orthorhombic), 2/m, C2/m.

$$Z = 2$$
$$á = 1.736 - 1.747$$
$$â = 1.742 - 1.753$$
$$ã = 1.748 - 1.761$$

Plane Polarized Light

Form and Relief: Staurolite occurs as squat prisms, usually containing inclusions of quartz and other matrix minerals. It occurs as porphyroblasts (Fig. 11.15A). Cross-sections with six sides are common. Relief is high.

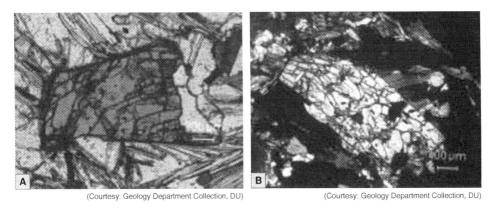

Fig. 11.15 Photomicrographs of a staurolite porphyroblast. **A.** Plane polarized light; **B.** Crossed polars.

Color and Pleochroism: Yellow or pale yellow; strongly pleochroic with á = colorless, â = pale yellow, ã = golden yellow.

Cleavage: {010} moderate to poor, seldom observed under microscope.

Under Crossed Polars

Anisotropism: Anisotropic.

Birefringence: Low but may be masked by yellow color of the mineral. Best seen in {100} sections.

ã: 0.011 – 0.014.

Polarization Colors: First order yellow or grey (Fig. 11.15B).

Extinction: Straight on prism edges or cleavage.

Orientation: á = y, â = x, ã = z (Fig. 11.16).

OAP: (100).

Indicatrix: Biaxial (+); {010} and {001} sections are suitable sections; weak birefringence causes practically no colored rings.

$2V_a$: 80° – 90°.

Sign of Elongation: Positive, length slow.

Dispersion: Weak, r > v.

Twinning: Rarely seen in thin sections.

Zoning: At times shows sector zoning; it is not infrequent and is sometimes defined by a chiastolite-like arrangement of inclusions

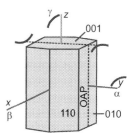

Fig. 11.16 Optic orientation of staurolite.

Alteration

Rare but may alter to green ferric chlorite.

Distinguishing Characters

Staurolite has a marked pleochroic scheme, with a typical yellow color and low birefringence that distinguish it from other minerals. It can be differentiated from vesuvianite by its biaxial nature

Mineral Association

Staurolite is a typical mineral of regional metamorphic rocks such as a metapelites. Associated minerals are garnet, biotite, feldspars, cordierite, kyanite etc.

Sphene

Nature of the Phase

Treated as pure phase with stoichiometric composition as $CaTi[SiO_4](O,OH,F)$. Al or Fe^{+3} may replace Si with paired substitution of one oxygen by (OH), or Na for Ca when rare earths substitute for Ti.

Crystallography

Monoclinic. 2/m, $P2_1/a$.

Z = 4
á = 1.843 – 1.950
â = 1.870 – 2.034
ã = 1.943 – 2.110

Plane Polarized Light

Form and Relief: Small to large commonly euhedral crystals exhibiting characteristic diamond shape. High relief (Fig. 11.17).

Color and Pleochroism: In general, colorless, when colored there is weak pleochroism. á = colorless or pale yellow, â = pale yellow to brown, ã = orange brown. Absorption formula: ã > â > á.

Cleavage: Good prismatic cleavage {110} but normally seen only in handspecimens.

(Courtesy: N.C. Pant)

Fig. 11.17 Photomicrograph of diamond shaped sphene.

Under Crossed Polars

Anisotropism: Anisotropic.

Birefringence: Strong. Best seen in {010} sections.

ä: 0.1 – 0.192.

Polarization Colors: Very high order of colors seen as pearl grey (Fig. 11.17).

Extinction: Inclined extinction. $x \wedge á = 6° – 20°$, $z \wedge ã = 36° – 51°$.

Orientation: $á \wedge x = 21°$, $â = y$, $ã \wedge z = 51°$ (Fig. 11.18).

OAP: (010).

Indicatrix: Biaxial (+) figure seen in basal sections. High ä allows a large number of colored rings which remain in the field of view during rotation of the stage since 2 V is small.

$2V_ä$: 17° – 56°.
$2V_ä$ decreases with the substitution of Fe^{3+} and Al for Ti.

Sign of Elongation: Not relevant as optic directions are at high inclination with crystallographic directions (Fig. 11.18).

Dispersion: Strong; r > v. Optic axis dispersion.

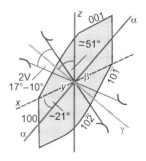

Fig. 11.18 Optic orientation of sphene.

Nesosilicates

Twinning: Simple twinning on {100} seen as twin plane trace along the long diagonal of the diamond shaped grains. In several instances, lamellar twinning on {221} is also seen.

Zoning: Seen in some grains.

Alteration

Common. The altered grain acquires an opaque character; it is commonly referred to as *leucoxene*. It is a fine-grained mixture of rutile, anatase, quartz, ilmenite etc.

Distinguishing Characters

Diamond shaped grains of extreme relief and high birefringence and high dispersion distinguish it from monazite, biaxial character distinguishes it from cassiterite, rutile, and xenotime. As a biaxial positive mineral it is different from biaxial negative baddeleyite.

Mineral Association

Very common in acidic plutonic rocks but abundant in intermediate rocks like syenite; in such rocks it occurs with pyroxenes and opaques.

Topaz

Nature of the Phase

A fluorine bearing silicate, occurs as a pure phase with a constant composition of $Al_2[SiO_4](OH,F)_2$.

Crystallography

Orthorhombic, 2/m2/m2/m, Pbnm.

$$Z = 4$$
$$á = 1.606 - 1.634$$
$$â = 1.609 - 1.637$$
$$ã = 1.616 - 1.644$$

Plane Polarized Light

Form and Relief: Usually occurs as prismatic crystals, subhedral to anhedral with a moderate to high relief. Cross-sections may be square, diamond-shaped, or eight sided. Often contains inclusions of opaque iron minerals, fluid inclusions also common.

Color and Pleochroism: In general, colorless, but thick sections may appear yellowish or pink in color. Thick sections show pleochroism with á = yellow, â = pale orange, ã = pink.

Cleavage: {001} perfect.

Under Crossed Polars

Anisotropism: Anisotropic.

Birefringence: Low, best seen in {010} sections.

ä: 0.008 – 0.011.

Polarization Colors: First order yellow or grey.

Extinction: Parallel.

Orientation: á = x, â = y, ã = z (Fig. 11.19).

OAP: (010).

Indicatrix: Biaxial (+), interference figure is easily obtained on basal sections (cleavage fragments). Isochromes not normally seen.

$2V_á$: 48° – 68°.
$2V_ã$ decreases as (OH) replaces F.

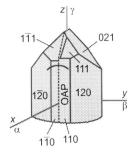

Fig. 11.19 Optic orientation of topaz.

Sign of Elongation: Positive elongation, length slow.

Dispersion: r > v.

Twinning: Rarely seen in thin sections

Zoning: Not seen.

Alteration

Topaz alters to clay minerals (kaolin) and sericite due to hydrothermal alteration

Distinguishing Characters

Weak birefringence, positive optic sign, prismatic form and moderate 2V are characteristic properties of topaz. It can be distinguished from andalusite by its smaller 2V and different orientation. It has higher birefringence than mellilite or vesuvianite. It may resemble quartz and feldspar in color and birefringence but the relatively high relief is distinctive.

Mineral Association

Quartz, microcline, muscovite, tourmaline etc. in granites, pegmatites, and pneumatolytic quartz veins.

Chloritoid

Nature of the Phase

$(Fe^{2+},Mg,Mn)_2(Al,Fe^{3+})Al_3O_2[SiO_4]_2(OH)_4$. Significant variations in Mg and Mn contents are common. Ca, Ti and some other elements may be present in minor amount. Mn-rich variety is not uncommon and was known by the name *ottrelite*.

Crystallography

Triclinic, $\bar{1}$, $C\bar{1}$. Monoclinic polytypes also common.

$$Z = 2$$
$$á = 1.705 - 1.730$$
$$â = 1.708 - 1.734$$
$$ã = 1.712 - 1.740$$

Plane Polarized Light

Form and Relief: Small elongated euhedral and anhedral porphyroblasts are common. Inclusions of quartz and other minerals are characteristically present. Nearly hexagonal basal sections. Platy habit (Fig. 11.20).

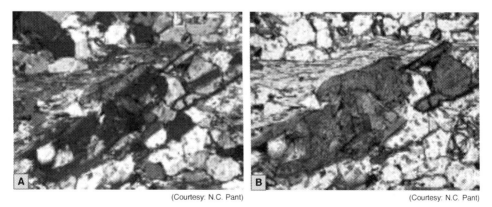

(Courtesy: N.C. Pant) (Courtesy: N.C. Pant)

Fig. 11.20 Photomicrographs of a porphyroblast of chloritoid. **A.** Plane polarized light; **B.** Crossed polar.

Color and Pleochroism: Green, pleochroism distinct. á = pale green, â = indigo blue, pale green, ã = colorless. Absorption Formula â > á > ã.

Cleavage: Perfect {001}, moderate {110}. In some case a parting parallel to {010} also seen.

Under Crossed Polars

Anisotropism: Anisotropic.

Birefringence: Weak to Moderate. Best seen in the {010} or near {100}.

ä: 0.005 – 0.022.

Polarization Colors: First order grey. Often anomalous.

Extinction: 10° – 18° in prismatic sections, symmetrical in basal sections.

Orientation: May vary from grain to grain; Monoclinic poly type may have y= â or á, z∧ ã = 2° – 30° (Fig. 11.21).

OAP: (010).

Indicatrix: It ranges from Biaxial (+) to Biaxial (–), though positive variety is more common. The best figure is seen in basal pinacoid.

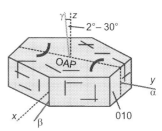

Fig. 11.21 Optic orientation of chloritoid.

$2V_{\text{ä}}$: 36° – 145°.

Dispersion: Strong; r > v bisectrix dispersion noticed.

Twinning: Commonly twinned on {001}, multiple twinning.

Zoning: Color zoning and hourglass structure common.

Alteration

May alter to chlorite or sericite.

Distinguishing Characters

High relief distinguishes it from chlorite. Twinning and hourglass zoning are other characteristic features.

Mineral Association

A regional metamorphic mineral in metapelites. Common associated minerals are quartz, chlorite, albite, garnet, muscovite, etc.

12
Sorosilicates and Cyclosilicates

- Zoisite
- Clinozoisite – Epidote Series
- Allanite
- Piemontite
- Pumpellyite
- Lawsonite
- Beryl
- Tourmaline
- Cordierite

Technical Text Box 12.1

When two silicate tetrahedra are linked with each other via a bridging oxygen an $[Si_2O_7]^{6-}$ oxyanion results. The presence of oxyanion does not exclude the possibility of independent silicate tetrahedra in the unit cell (look at the structure of epidote). The silicate group is named as the sorrosilcate group. This is a small group in terms of number of minerals.

When four or more silicate tetrahedra link in such a way so as to form rings with a general formula of $[Si_nO_{3n}]^{2n-}$. These silicates are called Cyclosilicates. Like sorosilicates this group also contains only a few important minerals.

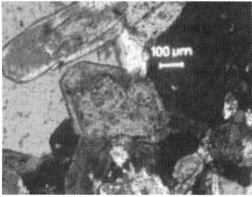

(Courtesy: P. Singh)

A photomicrograph of a euhedral epidote from the metamafites of Lake Morari, Ladakh.

Zoisite

Nature of the Phase

Nearly a pure phase. $Ca_2Al_2O.AlOH$ $[Si_2O_7]$ $[SiO_4]$. A limited solid solution occurs between zoisite and ferrian zoisite. These two zoisite are called á-zoisite and â-zoisite. The distinction is based on Fe content. Up to $X_{Fe}\left[=\dfrac{n(Fe)}{n(Fe+Al)}\right]=$ 0.03 it is á-, and for higher X_{Fe} content the variety is called â.

Crystallography

Orthorhombic, 2/m2/m2/m, Pnma.

$$Z = 4$$
$$á = 1.685 - 1.705$$
$$â = 1.688 - 1.710$$
$$ã = 1.697 - 1.725$$

Plane Polarized Light

Form and Relief: Usually found in clusters of elongated prismatic crystals, with rectangular cross-sections. Other common habits are bladed and fibrous. Relief of the mineral is high.

Color and Pleochroism: Usually colorless, a pink variety (thulite) may occur in Mn-rich environment. Thulite shows pleochroism with á = pale pink-dark pink, â = nearly colorless, ã = pale yellow.

Cleavage: {100} perfect, {001} imperfect. Cleavage fragments common.

Under Crossed Polars

Anisotropism: Anisotropic.

Birefringence: Very low (varies between $0.003 - 0.008$); maximum birefringence seen on (010) section for ferrian zoisite, on (100) section for zoisite.

ä: $0.003 - 0.008$.

Polarization Colors: First order grey color in á-zoisite, but â-zoisite shows anomalous deep berlin blue color.

Extinction: Straight on the prism edge or {100} cleavage.

Orientation: á = x, â = y and ã = z (á-zoisite) and á = y, â = x and ã = z (â-zoisite).

OAP: (010) for á-zoisite and (100) for â-zoisite (Fig. 12.1).

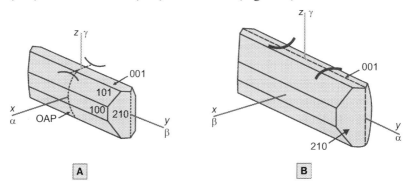

Fig. 12.1 A. Optic orientation of zoisite; B. Optic orientation of ferrian zoisite.

Indicatrix: Biaxial (+); cleavage fragments would yield flash figure for á-zoisite, but an obtuse bisectrix for â-zoisite.

$2V_{\mathrm{\acute{a}}}$: $0° - 69°$.
The most common $2V_{\mathrm{\~a}}$ is 30 for á-zoisite but can even be 0, (i.e., uniaxial) in some cases; becomes equal to 60.

Dispersion: Strong. r < v (á-zoisite) and r > v (for â-zoisite).

Sign of Elongation: Cleavage traces in á-zoisite are fast direction, however, in the case of â-zoisite, these could either be length fast or slow. Zoned grains, often, exhibit contrasting sign of elongation in adjacent zones.

Twinning: Not common.

Zoning: Composition zoning (ferric iron content) is common; reflected in 2V variation. In thulite color zoning has been observed.

Alteration

None.

Distinguishing Characters

Can be distinguished from clinozoisite by parallel extinction and from epidote by lack of color (except thulite). Vesuvianite is biaxial.

Mineral Association

A metamorphic mineral characteristically associated with albite, calcite, quartz, garnet, calcic amphibole, etc.

Clinozoisite – Epidote Series

Nature of the Phase

A binary solid solution series $CaAl_2O.AlOH[Si_2O_7][SiO_4] - CaAl_2O.(AlFe)OH[Si_2O_7][SiO_4]$. While the iron-free end-member, *i.e.*, clinozoisite is very common in nature, iron-end-member occurs only as synthetic product. In nature about one atom per formula unit Fe^{3+} is the maximum reported substitution. The molecule $Ca_2Fe^{3+}Al_2(SiO_4)_3 (OH)$ is called the pistacite molecule (Ps). The suggested change of name from clinozoisite to epidote is when the optic character changes from (+) to (–); it occurs at about 6wt% Fe_2O_3 (or ~ Ps_{10}), though higher values have also been reported.

Crystallography

Monoclinic, 2/m, $P2_1/m$.

$Z = 2$		Clinozoisite	Epidote
á	=	1.670 – 1.718	1.715 – 1.751
â	=	1.670 – 1.725	1.725 – 1.784
ã	=	1.690 – 1.734	1.734 – 1.797

The refractive indices increase as Fe^{3+} progressively replace Al.

Plane Polarized Light

Form and Relief: Found in granular aggregates, which are usually quite small, with a high relief (Fig. 12.2). When it occurs as euhedral, crystals may be eight sided.

Color and Pleochroism: Colorless, may be pale yellow and is non pleochroic for clinozoisite composition. Epidote distinctly pleochroic. á = colorless, â = yellow-green, ã = colorless to green. Absorption formula: â > á > ã.

Cleavage: Basal cleavage {001} perfect. It appears as a prismatic cleavage in sections, since the mineral is elongated parallel to *b*-axis.

Under Crossed Polars

Anisotropism: Anisotropic.

Birefringence: Very low for the clinozoisite end-member. Increases with Fe content; The rate of change of ä abruptly increases for Ps > 15%. Epidote birefringence is moderate to strong.

ä: 0.004 – 0.015 for clinozoisite; 0.0012 – 0.049 for epidote.

Sorosilicates and Cyclosilicates

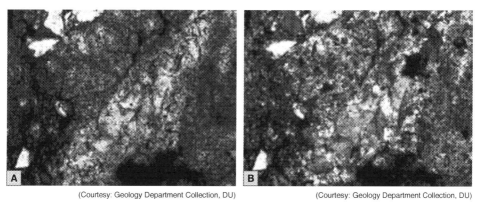

Fig. 12.2 Photomicrographs of Clinozoisite grains. **A.** In an amphibolite, plane polarized light ; **B.** Crossed polars; **C.** An eclogite from the Himalayas.

Polarization Colors: First order grey for the clinozoisite often, anomalous due to strong absorption. Epidote shows high second order colors up to second order blue.

Extinction: Oblique, extinction angles varies depending on the composition. Most elongate prismatic sections have straight extinctions; in non-prismatic sections the angle may show variation from 0° – 60°.

Orientation: It is strongly composition dependent as shown below:

	Clinozoisite	Epidote
$á \wedge z$ =	−85° – 0°	0° – 7°
$â$ =	y	y
$ã \wedge x$ =	−60° – 25°	25° – 45°

OAP: (010) for all compositions. *See* Figs. 12.3A and B for orientation.

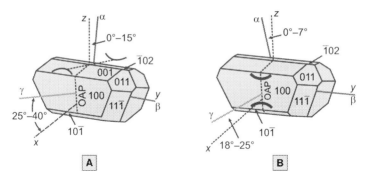

Fig. 12.3 A. Optic orientation of clinozoisite; **B.** Optic orientation of Epidote.

Indicatrix: Clinozoisite   Epidote
              Biaxial (+)   Biaxial (–)

$2V_{\hat{a}}$: Clinozoisite   Epidote
              14° – 90°   90° – 155°

The value of $2V_{\hat{a}}$ is dependent on Fe^{3+} content. However, the relationship between $2V_{\hat{a}}$ and Fe^{3+} content does not have a straight-line relationship. Optic figure is not easily seen because of small grain size habit. In coarse grain crystals (100) is best suited for clinozoisite optic figure; in the case of epidote basal pinacoid exhibits acute bisectrix figure. However, because of composition dependence of optic orientation and zoing this prediction may not be true for all grains.

Dispersion: $r < v$ strong $r > v$ strong.

Sign of Elongation: Cleavage traces may be length slow or length fast since $\hat{a}$ is parallel to y-direction.

Twinning: Not common, rare on {100}.

Zoning: Compositional zoning is very common in most natural examples. It is frequently expressed as color zoning, or in the variation of extinction angle, 2V, birefringence etc.

Alteration

None.

Distinguishing Characters

Clinozoisite has lower birefringence, optically positive character and lack of pleochroism in comparison to epidote. From zoisite it is distinguished by its oblique extinction in prism zone sections, and from mellilite and vesuvianite by its biaxial character. These latter minerals are of low birefringence as compared to epidote.

Sorosilicates and Cyclosilicates

Mineral Association

The minerals of this series owing to exsolution within their solid solution commonly may be found together, particularly if the Fe content and the oxygen fugacity are favourable. Occurs in marbles and amphibolites. The minerals associated are calcite, albite, chlorite, calc-amphiboles, garnet etc. Also common in granites crystallized at comparatively higher pressures.

Allanite (Orthite)

Nature of the Phase

A multi-component solid solution as reflected in the following accepted formula:

$$(Ca, Mn, Ce, La, Y, Th)_2(Fe^{2+}, Fe^{3+}, Ti)(Al, Fe^{3+})_2 O.OH[Si_2O_7][SiO]_4$$

A host of elements of the rare earth group as well as other have elements occur in the Ca sites, bivalent metals replace the trivalent ions in the octahedral site.

Crystallography

Monoclinic, 2/m, $P2_1/m$.

$Z = 2$
$á = 1.690 – 1.813$
$â = 1.700 – 1.857$
$ã = 1.706 – 1.891$

The wide range in the values of refractive index is a reflection of its complicated chemical composition.

Plane Polarized Light

Form and Relief: Common habit is euhedral and acicular. In some cases it forms a core of a zoned epidote. Relief high (Fig. 12.4).

Color and Pleochroism: Light brown. Pleochroic in brown; á = colorless to light brown, â = pale to red brown, ã = dark brown. Rarely pleochroic in green to brown black. ã > â > á.

Cleavage: Basal cleavage is perfect {001}; {100} and {110} poorly seen.

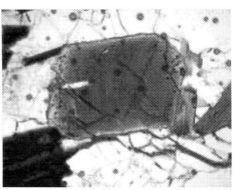

(Courtesy: A. Kundu)

Fig. 12.4 A zoned allanite grain in plane polarized light.

Anisotropism: Anisotropic.

Birefringence: Low to moderate. Maximum birefringence is seen on (010).

δ: 0.013 – 0.036.

Polarization Colors: Second order colors, though invariably masked by original intense color. May also occur as metamict.

Extinction: Crystals cut parallel to b show parallel extinction; other sections show inclined extinction with extinction angle measuring up to 40°.

Orientation: $\alpha \wedge z = 4° - 47°$, $\beta = y$, $\gamma \wedge x = 26° - 72°$ (Fig. 12.5).

OAP: (010).

Indicatrix: Biaxial (+) and also (–), negative allanite is common. Very difficult to obtain optic figure because of zoning and metamict nature.

2V$_\alpha$: 40° – 123°.

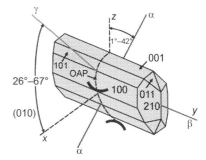

Fig. 12.5 Optic orientation of allanite.

Sign of Elongation: Can be length fast or slow since β coincides with y.

Dispersion: Moderate to strong, $r > v$.

Twinning: Not common, rare on {001} and {100}.

Zoning: Zoning commonly seen; thin margins often are epidote.

Alteration

Commonly occurs as metamict.

Distinguishing Characters

Can be distinguished from other epidote group of minerals by its yellow to reddish brown color is distinct. In comparison to brown biotite and brown hornblende it has poor cleavage.

Mineral Association

It is an accessory mineral in granites and pegmatites. Generally associated with other radioactive and rare earth bearing minerals, also with epidote

Piemontite

Nature of the Phase

Essentially a Mn-epidote but is characterized by a large variation in composition from $Ca_2[Fe^{3+}, Al]_3O.OH[Si_2O_7][SiO_4]$ to $Ca_2[Mn^{3+}, Al]_3O.OH[Si_2O_7][SiO_4]$ to $Ca_2[Mn^{3+}_2Al]O.OH[Si_2O_7][SiO_4]$. General formula: $Ca_2[Mn^{3+}, Fe^{3+}, Al]_3O.OH[Si_2O_7][SiO_4]$.

Crystallography

Monoclinic, 2/m, $P2_1/m$.

$Z = 2$
$á = 1.730 - 1.794$
$â = 1.740 - 1.807$
$ã = 1.762 - 1.829$

Plane Polarized Light

Form and Relief: Common habit is radial aggregates of acicular crystals. Cross-sections varied: rectangular, hexagonal, or octangular.

Relief: High.

Color and Pleochroism: Brown. Strongly pleochroic in brown to red. á = pale yellow, â = violet to red-violet, ã = deep red.

Cleavage: {001} perfect; cleavage fragments common.

Under Crossed Polars

Anisotropism: Anisotropic.

Birefringence: Moderate to strong. Maximum birefringence is seen on (010).

ä: $0.025 - 0.082$.

Polarization Colors: Second to even third order colors, though invariable masked by original intense color.

Extinction: Best seen in (010) section with extinction angle up to 35°. Parallel extinction along the length of long grains.

Orientation: á ∧ z = 2° – 9°, â = y, ã ∧ x = 27° – 35° (Fig. 12.6).

OAP: (010).

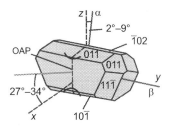

Fig. 12.6 Optic orientation of piemontite.

230 Optical Mineralogy

Indicatrix: Biaxial (+) interference figure seen on (100); may be masked by strong absorption.

$2V_a$: $50° – 86°$.

Sign of Elongation: Cleavage traces are length fast.

Dispersion: Strong, $r > v$. Optic Axis dispersion.

Twinning: Rare lamellar twinning on {100}.

Zoning: Zoning commonly seen as color and optic constants change vastly.

Alteration

Not common.

Distinguishing Characters

Strong pleochroism makes piemontite identification easy.

Mineral Association

Low grade metamorphic mineral; commonly occurring with quartz, sericite, chlorite etc.

Pumpellyite

Nature of the Phase

A limited solid solution between Fe^{3+} Al exists in pumpellyite. Besides this substitution, Fe^{2+}, Mn and Cr may also substitute in minor amount. The general formula is $Ca_2Al_2(Al, Fe^{2+}, Mg, Fe^{3+})[Si_2(O,OH)_7][SiO_4](OH, O)_3$.

Crystallography

Monoclinic. 2/m. A2/m.

$$Z = 4$$
$$á = 1.665 – 1.710$$
$$â = 1.675 – 1.720$$
$$ã = 1.683 – 1.726$$

When subsitution of Fe, Mn and Cr increases, refractive index values higher than reported above may also be found.

Plane Polarized Light

Form and Relief: Acicular and rosette grains very common, in some instances also bladed. Commonly subhedral or anhedral. Relief moderate to strong.

Color and Pleochroism: Green; strongly pleochroic. Pleochroic Scheme: á = colorless to pale green, â = blue to deep green, ã = colorless to yellow or reddish brown. â > ã ≥ á.

Cleavage: Two sets; distinct on (001), less clear (100).

Under Crossed Polars

Anisotropism: Anisotropic.

Birefringence: Low, best seen in (010) sections. Subsitution of Fe^{2+}, Mn, etc., increases birefringence.

ä: 0.002 – 0.022.

Polarization Colors: Low first order yellow to red.

Extinction: Usually parallel particularly on sections elongated on b-axis. Normal to b-axix sections show 0° – 34° extinction angle with traces of (001) cleavage.

Orientation: Highly uncertain. In majority of cases y = â but there are examples of y = ã. á ∧ x = 4° – 32°, ã ∧ z = –4° – +34°. (Fig. 12.7).

Indicatrix: Biaxial (+) as well as biaxial (–).

$2V_a$: Highly variable. 7° – 110°. Up to 150° in Fe-rich variety.

Sign of Elongation: Cleavage traces show positive and negative elongation in different examples. Fe-rich pumpellyite is commonly length slow.

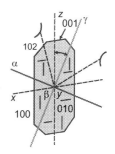

Fig. 12.7 Optic orientation of pumpellyite.

Dispersion: Strong, particularly of optic axis.

Twinning: Twinning (001) on and (100) planes very common; because of twining the grains are divided into four-fold sectors.

Zoning: Zoning commonly seen as dark color rim indicating Fe enrichment.

Alteration

Not common.

Distinguishing Characters

Its pleochroism is characteristic. Has higher relief and birefringence than chlorite.

Mineral Association

Low grade metamorphic mineral commonly occurring with quartz, prehnite, epidote etc.

Lawsonite

Nature of the Phase

Almost a pure phase: $CaAl_2Si_2O_7(OH)_2$. H_2O with very minor substitution of Fe3+, Ti etc.

Crystallography

Orthorhombic, 2/m2/m2/m, Ccmm.

$$Z = 4$$
$$á = 1.665$$
$$â = 1.672 - 1.676$$
$$ã = 1.684 - 1.686$$

Plane Polarized Light

Form and Relief: Tabular crystal on (010), commonly euhedral. Relief moderate.

Color and Pleochroism: Colorless. If a section of the mineral is thicker than 0.03mm then it may show yellow color with pleochroism from yellow to blue. á = blue â = yellow ã = colorless. Absorption Formula á > â > ã.

Cleavage: Several sets of cleavages are present. Perfect along {010}, good along {100}, faint along {101}. The first two intersect at nearly 90°; last one appearing as two sets due to the presence of mirror planes, intersect at 67°.

Under Crossed Polars

Anisotropism: Anisotropic.

Birefringence: Moderate, best seen in (010) section.

ä: 0.019 – 0.021.

Polarization Colors: Middle first order colors up to red.

Extinction: Parallel extinction with {100} and {010} cleavages as well as length of elongated crystals.

Orientation: á = z, â = x, ã = y (Fig. 12.8).

OAP: {1010}.

Indicatrix: Biaxial (+). Acute bisectrix figure best seen on (010).

$2V_a$: 76° – 87°.

Sign of Elongation: Negative elongation, i.e., length fast with reference to cleavage {010} and {100}. Accicular grains extending along y-direction are length slow.

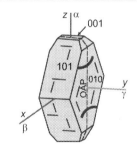

Fig. 12.8 Optic orientation of lawsonite

Dispersion: Very strong, r > v.

Twinning: Twinning is common. Both simple and multiple twins are present on {110}.

Zoning: Not present.

Distinguishing Characters

Higher birefringence distinguishes it from clinozoisite. Lawsonite has lower birefringence as compared to prehnite.

Alteration

Commonly no alteration.

Mineral Association

Low temperature metamorphic mineral commonly occurring with quartz; characteristic mineral in hydrothermal veins as an alteration product of calcic-plagioclase. Also, a characteristic mineral of blue schist where it occurs with glaucophane.

Beryl

Nature of the Phase

Treated as a pure phase with stoichiometric composition of $Be_3Al_2[Si_6O_{18}]$. Many natural grains invariably contain some alkalis, i.e., Na, Li, K and Cs.

Crystallography

Hexagonal, 6/m2/m2/m, P6/mcc.

Z = 2
ù = 1.569 – 1.610
â = 1.565 – 1.610

Plane Polarized Light

Form and Relief: Usually occurs as hexagonal prisms or more commonly as anhedral grains under microscope. Relief is low to moderate (Figs. 12.9A and B).

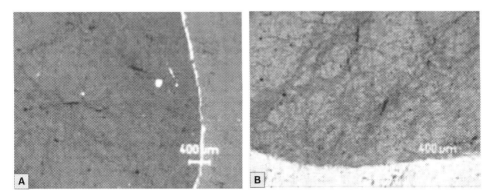

Fig. 12.9 Photomicrographs of beryl. **A.** Plane polarized light; **B.** Crossed polars.

Color and Pleochroism: Color of beryl varies from colorless, pale yellow to pale green and show pleochrism in thick sections in pale green. Pleochroic Scheme: ù = yellowish green, å = sea green.

Cleavage: Cleavages are imperfect on {0001}.

Under Crossed Polars

Anisotropism: Anisotropic.

Birefringence: Low, similar to quartz and feldspar.

ä: $0.004 - 0.009$.

Polarization Colors: Low first order grey.

Extinction: Parallel, though in the absence of any crystal marker not easy to determine.

Indicatrix: Uniaxial (−). Interference figure consists of broad dark isogyres with a grey background.

Sign of Elongation: Length fast.

Dispersion: Weak.

Twinning: Rarely present.

Zoning: Zoning is common, owing to different structural states, ionic substitutions, and common inclusions. It may appear as reaction rims over kyanite; quartz rims over cordierite are known.

Sorosilicates and Cyclosilicates   235

Alteration

Kaolinite and muscovite are common alteration product.

Distinguishing Characters

Beryl can be confused with quartz but can be distinguished by its higher refractive indices and length fast orientation.

Mineral Association

Occurs in granites and pegmatites with quartz, feldspars etc.

Tourmaline

Nature of the Phase

A boron bearing silicate with complex chemistry. The general formula is $XY_3Z_6[T_6O_{18}][BO_3]_3V_3W$. Where X = Ca, Na, K, □; Y = Li, Mg, Fe^{2+}, Mn^{2+}, Al, Cr^{3+}, V^{3+}, Fe^{3+}, (Ti^{4+}); Z = Mg, Al, Fe^{3+}, V^{3+}, Cr^{3+}; T = Si, Al, (B); B = B, (□); V = OH, O = [O(3)]; W = OH, F, O = [O(1)].[1] Some important end-members are Dravite $(NaMg_3Al_6B_3Si_6(O, OH)_{30}(OH, F)$, Schorl $(Na(Fe,Mn)_3Al_6B_3Si_6(O, OH)_{30}(OH, F)$ and Elbaite $(Na(Li, Al)_3Al_6B_3Si_6(O, OH)_{30}(OH, F)$. A partial list is given in the following table:

Important tourmaline end-member species

Species	(X)	(Y_3)	(Z_6)	T_6O_{18}	$(BO_3)_3$	V_3	W
Alkali tourmaline							
Elbaite	Na	$Li_{1.5}Al_{1.5}$	Al_6	Si_6O_{18}	$(BO_3)_3$	$(OH)_3$	(OH)
Dravite	Na	Mg_3	Al_6	Si_6O_{18}	$(BO_3)_3$	$(OH)_3$	(OH)
Chromdravite	Na	Mg_3	Cr_6	Si_6O_{18}	$(BO_3)_3$	$(OH)_3$	(OH)
Schorl	Na	Fe^{2+}_3	Al_6	Si_6O_{18}	$(BO_3)_3$	$(OH)_3$	(OH)
Calcic tourmaline							
Uvite	Ca	Mg_3	Al_5Mg	Si_6O_{18}	$(BO_3)_3$	$(OH)_3$	F
"Hydroxy-feruvite"	Ca	Fe^{2+}_3	Al_5Mg	Si_6O_{18}	$(BO_3)_3$	$(OH)_3$	(OH)
X-site-vacant tourmaline							
Rossmanite		$LiAl_2$	Al_6	Si_6O_{18}	$(BO_3)_3$	$(OH)_3$	(OH)
Foitite		Fe^{2+}_2Al	Al_6	Si_6O_{18}	$(BO_3)_3$	$(OH)_3$	(OH)
Magnesiofoitite		Mg_2Al	Al_6	Si_6O_{18}	$(BO_3)_3$	$(OH)_3$	(OH)

1. Hawthorne, F.C. and Henry, D.J. 1999. Classification of the minerals of the tourmaline group. Eur. J. Mineral. 11. 201-215.

Optical Mineralogy

Crystallography

Trigonal (rhombohedral-Hexagonal), 3m, R3m.

Z = 3	Dravite	Schorl	Elbaite
$\grave{u}$ =	1.612 – 1.632	1.635 – 1.650	1.615 – 1.630
$\mathring{a}$ =	1.634 – 1.661	1.660 – 1.671	1.633 – 1.651
$\ddot{a}$ =	0.021 – 0.029	0.025 – 0.035	0.017 – 0.021

The indices increase with increasing amount of (Fe^{2+}, Fe^{3+}, Mn, Ti).

Plane Polarized Light

Form and Relief: Usually occurs as large elongate prismatic crystals, often occurring in radiating clusters. Crystal cross sections may be six sided but commonly trigonal. Relief is moderate (Chapter XX, Fig. YY).

Color and Pleochroism: Color is highly variable and often irregular from colorless to blue, green or yellow depending on the composition. Fe bearing tourmaline *i.e.*, Schorl are darker in sheds from brown to black. Elbaites tend to be of light shades of blue, pink, green or colorless and dravite vary from dark brown to pale yellow. The pleochroism is variable in intensity but is particularly strong for Fe bearing tourmalines.

Pleochroic Scheme: $\grave{u}$ is always greater than $\mathring{a}$

         Dravite $\grave{u}$ = dark brown $\mathring{a}$ = pale yellow

         Schorl $\grave{u}$ = dark green $\mathring{a}$ = redish violet

         Elbaite $\grave{u}$ = blue $\mathring{a}$ = pale green, pale yellow

Due to high intensity of color, absorption formula is not easy to detect. Since $\grave{u}$ is normal to z direction which when placed along E-W cross wire under plane polarized light will enable the grain to absorb maximum amount of light. This property has been in connection with checking of the orientation of planes of vibration of the lower polar.

Cleavage: Imperfect to poor prismatic {120} and pyramidal {101} cleavages. Fractures perpendicular to c {0001} are common.

Under Crossed Polars

Anisotropism: Anisotropic.

Birefringence: Strong, colors of second order are seen, Fe bearing variety shows the strongest birefringence with. Best seen in prismatic sections.

ä: 0.19 – 0.35.

Polarization Colors: Second order colors; interference colors up to third order in high Fe varieties.

Extinction: Gives straight extinction along the prism edge.

Indicatrix: Uniaxial (–) seen on the basal section with a few isochrones. High absorption varieties may not show the figure.

Dispersion: Weak $r > v$.

Twinning: Rare on {101}.

Zoning: Concentric color zoning is common. Outer zones tend to be darker. Best seen in basal section.

Alteration

Alteration product of tourmaline include muscovite, biotite or lepidolite, and also chlorite and cookeite, $(LiAl_4(Si, Al)_4O_{10}(OH)_8)$.

Distinguishing Characters

Strong pleochroism is most characteristic property of all varieties of tourmaline. Biotite and hornblende display perfect cleavages parallel to probable elongation and show greatest absorption of light vibrating parallel to elongation. Distinction between different varieties of tourmaline is referred to by the refractive indices or birefringence.

Mineral Association

A characteristic association of granites and pegmatites. It may crystallize with the rock or may form along fractures within the hydrothermal veins. Most characteristic mineral associated with tourmaline is quartz.

Cordierite

Nature of the Phase

Limited binary solid solution between Mg and Fe rich varieties. $Al_3(Mg, Fe)_2(Si_5Al_4O_{18}).nH_2O$.

Crystallography

Orthorhombic, 2/m2/m2/m, Cccm.

$$Z = 4$$
$$á = 1.527 - 1.560$$
$$â = 1.532 - 1.574$$
$$ã = 1.577 - 1.578$$

Variation in the Fe and H_2O content of the mineral would change the indices. The values decrease with progressing (Fe + Mn) substitution for Mg. The relationship is complicated owing to structural complications of channels present in the structure. H_2O substitution increases values of indices.

Plane Polarized Light

Form and Relief: Pseudohexagonal cross-section grains as clusters, also anhedral. Relief poor (Fig. 12.11), may be negative or positive.

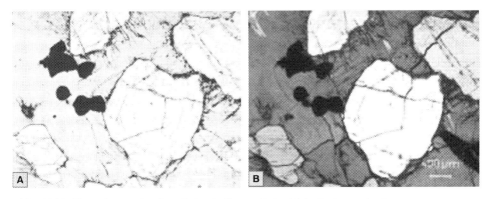

Fig. 12.11 Photomicrographs of cordierite. **A.** Plane polarized light. **B.** Crossed polars.

Color and Pleochroism: Colorless. Some Fe rich varieties may show pale color and pleochroism. á = colorless â = pale blue ã = blue; Absorption formula: ã > â > á. Pleochroic halos common, inclusion minerals are zircon, sphene, apatite and rutile giving out radiation.

Cleavage: Several sets of poor cleavages are present. Faintly seen parallel to {010}, others difficult to see.

Under Crossed Polars

Anisotropism: Anisotropic.

Birefringence: Very weak, best seen in (010) section.

ä: 0.019 – 0.021.

Polarization Colors: Middle first order colors up to red.

Extinction: Parallel extinction with all cleavages, but make 30° or 60° angle with twin planes.

Orientation: á = z, â = y, ã = x (Fig. 12.12).

OAP: {100}.

Indicatrix: Biaxial (–). Acute bisectrix figure best seen on (001). Flash figure on (010). Some varieties may be biaxial (+) also.

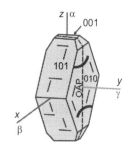

Fig. 12.12 Optic orientation of cordierite.

$2V_a$: 35° – 106°; most common range is 65° – 85°. (Fe + Mn) content of the grain content has profound effect on the value of $2V_a$, though even Be, Na, H_2O, CO_2 contents also affect the value of optic axial angle.

Sign of Elongation: Elongation not easily observable as grains are rarely euhedral and cleavages are not easily seen.

Dispersion: Weak, v > r. Optic axis dispersion.

Twinning: Twinning is very common. It may be simple, lamellar and cyclic. When cyclic it may radiate from centre outward with adjacent twin plane angles either 30°, 60°, or 120°. There are concentric twins also. In addition there are complex twins as well.

Zoning: Very common. In addition to compositional zoning, it may show zoning in the arrangement of inclusions. Twining causes sector-zoning.

Alteration

Easily altered to chlorite, muscovite and other hydrous magnesium minerals forming a fine-grained clay like mixture earlier termed pinite.

Distinguishing Characters

Easily confused with quartz from which it can be distinguished by its biaxial character. Sector (zoning) twinning whenever present may also make it distinct. Presence of pleochroic haloes around zircon and apatite inclusions may also help it identification.

Mineral Association

It is a common metamorphic mineral occurring within aureoles as well as in regional metamorphic terrains. Mica, plagioclase orthoclase, quartz garnet, biotite and chlorite are common associated.

■■■

13

Inosilicates

- Orthopyroxene Series
- Pigeonite
- Diopside – Hedenbergite Series
- Augite
- Omphacite
- Jadeite
- Aegirine to Aegirine – Augite
- Wollastonite
- Sapphirine
- Anthophyllite
- Cummingtonite – Grunerite Series
- Tremolite – Actinolite Series
- Calcic – Amphibole (Common Hornblende)
- Glaucophane
- Riebeckite

Technical Text Box 13.1

Inosilicates are also known as chain silicates because of the corner-sharing chains of silicate tetrahedra along z-direction. Apices of adjacent tetrahedra of each chain point to the opposite directions. Nature of these chains has given rise to a broad four-fold division of inosilicates. When chains are straight the minerals belong to pyroxene group; when these are not straight, then minerals are classified as pyroxenoids; and, when chains occur in pairs by bridging the third oxygens (double chains) then the minerals are called the amphiboles. The fourth group is known as bipyriboles where the unit cell structure is a combination of structures of pyroxenes, amphiboles, and sheet silicates.

These structural peculiarities allow a large number of metal and non-metal ions to occupy various structural sites giving rise to large number of minerals. Many of these minerals are important rock-forming minerals in the earth as well as in other planets and meteorites.

(Courtesy: Mineralogical Society of America)

A photomicrograph of jimthompsonite. Konishi, H., Buseck, P.R., Xu, H, Li, X. 2008. Protopolymorphs of jimthompsonite and chesterite in contact – metamophosed serpentinites from Japan, American Min., 93, 351-359.

© The Author(s) 2023
P. K. Verma, *Optical Mineralogy*,
https://doi.org/10.1007/978-3-031-40765-9_13

Orthopyroxenes Series

Nature of the Phase

Essentially a binary solid solution series between enstatite (En), $Mg_2(SiO_3)_2$ and orthoferrosilite (Fs), $Fe_2^{2+}(SiO_3)_2$.

Crystallography

Orthorhombic, 2/m2/m2/m, Pbca.

	Enstatite (En)	Orthoferrosilite (Fs)
á =	1.650	1.768
â =	1.653	1.770
ã =	1.658	1.788

Z = 16

The refractive indices have straight line relationships with total Fe(+Mn) content of orthopyroxene (Fig. 13.1). Note that á and ã vary with different slopes. ã is easily measurable value because of tabular habit of cleavage fragments which contain ã.

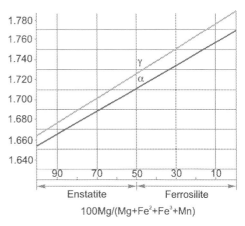

Fig. 13.1 The relationship of refractive indices to the Mg content in orthopyroxenes.

Plane Polarized Light

Form and Relief: Usually occurs as short, prismatic crystals with square to octagonal cross section (Figs. 13.2A and B).

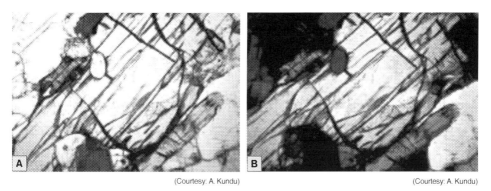

(Courtesy: A. Kundu)   (Courtesy: A. Kundu)

Fig. 13.2 Photomicrographs of an orthopyroxene grain. **A.** Plane polarized light; **B.** Crossed Polars.

Inosilicates

Color and Pleochroism: The color of orthopyroxene depends upon the ferrous content in it. End-member enstatite is colorless. Commonly, the presence of 10% ferrosilite mole imparts a distinct green color. Pleochroism weak initially but becomes progressively strong as Fe increases, though it is unclear whether the relationship is linear. When colored the mineral is pleochroic with á = pink, pale brown; â = pale brown, yellow; and ã = pale green, bluish green (Figs. 13.2A and B).

Cleavage: Excellent prismatic {210} cleavages; (210) ∧ ($\bar{2}$10) ~ 90°, partings {010} and {100} are also seen.

Under Crossed Polars

Anisotropism: Anisotropic.

Birefringence: Low, best seen on (010) section. It increases monotonically with increasing orthoferrosilite content (Fig. 13.3).

δ: 0.008 (Enstatite) – 0.022 (Ortho-ferrosilite).

Polarization Colors: Interference colors range from first order grey (enstatite) to yellow and red (iron rich members).

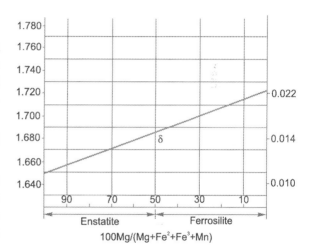

Fig. 13.3 The relationship of birefringence to the Mg content in orthopyroxenes.

Extinction: Parallel extinction on prism faces, but symmetrical on basal sections.

Orientation: α = y, β = x, γ = z (Fig. 13.4).

OAP: {010}. It is rarely shown as being parallel to {100}.

Indicatrix: Biaxial (+) for Mg rich members. Increasing ferrosilite substitution converts the indicatrix to biaxial (–) as shown in Fig. 13.5. Optic figure best seen in the basal section for enstatite composition.

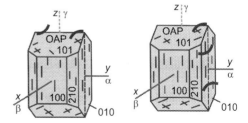

Fig. 13.4 Optic orientation of orthopyroxenes.

2V$_a$: Varies from 125° – 53° – 125°. The variation though related to X$_{Fe}$ content of orthopyroxene varies with a peak value at Fs$_{50}$ (Fig. 13.5).

Sign of Elongation: Cleavage traces on prismatic sections are positive (length slow).

Dispersion: It is also controlled by Fe content of the mineral though the relationship is ill-understood. Varies from weak to strong. Measured from á, it changes from $v > r$ to $r > v$ at En_{85} and back again to $v > r$ at En_{50}. Measured on acute bisectrix, $r > v$ for enstatite, then becoming $v > r$.

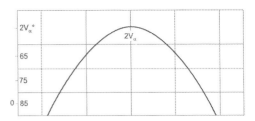

Fig. 13.5 The relationship of 2V to the Mg content in orthopyroxenes.

Twinning: Absent.

Zoning: Orthopyroxene phenocrysts occurring in volcanic rocks may show zoning with a Mg-rich core and Fe rich rim. Differential color zoning under plane polarized light shows it well.

Lamellar Structure: Orthopyroxenes show lamellar structure so characteristically that a special mention is necessary. There is a crystallographic control on the manner of lamellae. The lamellae lie along {100} plane of the host orthopyroxene and consist of Ca-rich clinopyroxene. As you will learn later in this chapter, the clinopyroxenes have comparatively high birefringence, hence, are conspicuous under cross polars giving rise to lamellar structure. This is considered to be an exsolution texture. Another lamellar structure is formed in Fe-rich orthopyroxene by inverted pigeonite lamellae that are aligned parallel to {001} of original pigeonite. In some cases lamellar structure is the result of glide dislocation in orthopyroxenes in {100} plane with glide direction {100} (Fig. 13.6).

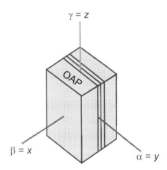

Fig. 13.6 A Sketch of host - lamellae relationship in orthopyroxene grains.

Alteration

Orthopyroxene may alter to serpentine; when the alteration is high enough that original orthopyroxene is seen with difficulty the grain is termed bastite. At times, it may alter around its rim to an amphibole (cummingtonite).

Distinguishing Characters

It is distinguished from other clinopyroxenes by their parallel extinction; from andalusite by its length slow character and from silimanite by lower order of interference colors. Another characteristic feature of many orthopyroxenes, particularly those moderately rich in iron is pleochroism from pink to green.

Inosilicates

Mineral Association

Orthopyroxenes are characteristic minerals of basic and ultrabasic plutonic and volcanic rocks. Its common associate minerals are augite and calcic plagioclase. It is also found in many granulites such as charnockites where it occurs with quartz, K-feldspar, plagioclase etc. It occurs with cordierite and microcline in hornfelses.

Pigeonite

Nature of the Phase

Essentially a binary solid solution involving Mg and Fe^{2+} exchange. Composition is close to that orthopyroxenes but with a small amount of Ca that ranges 0.05 to 0.15 X_{Ca}. (Mg, Fe, Ca)(Mg, Fe^{2+})[Si_2O_6]

Crystallography

Monoclinic, 2/m, $P2_1/c$.

$$Z = 4$$
$$á = 1.682 - 1.722$$
$$â = 1.684 - 1.722$$
$$ã = 1.705 - 1.751$$

Plane Polarized Light

Form and Relief: Grains elongated on c, octagonal outline.

Color and Pleochroism: Normally colorless; Fe rich varieties show pale green or brown color but pleochroism is rarely seen; maximum absorption on â.

Cleavage: Prismatic {110} cleavages; planes as two sets intersecting at 87° and 93°. Exsolution lamellae on {001} may appear as parting.

Under Crossed Polars

Anisotropism: Anisotropic.

Birefringence: Moderate to strong; maximum appears in {010} sections. Fe content of pigeonite increases the birefringence.

ã: 0.023 – 0.029.

Polarization Colors: Maximum upto middle second order.

Extinction: Maximum extinction angle range is 44° – 37° to cleavage traces, seen in {010} sections; symmetrical in basal sections {001}.

Orientation: Optic orientation in pigeonite is controlled by XCa and XFe values of the composition. For Ca and Fe poor grains $\alpha \wedge y = -19°$ to $-26°$, $\beta = x$, $\gamma \wedge z = -37°$ to $-44°$.

OAP: OAP is perpendicular to {010}. For comparatively Ca and Fe rich grains $\alpha \wedge x = -22°$ to $-26°$, $\beta = y$, $\gamma \wedge z = -40°$ to $-44°$. OAP is parallel to {010} (Fig. 13.7).

Indicatrix: Biaxial (+); may appear as nearly uniaxial. The isogyres are not easy to observe; {100} section in some cases and {001} in others may give an off-centred figure. The separation of isogyres is not apparent in most cases.

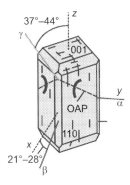

Fig. 13.7 Optic Orientation of pigeonite.

$2V_a$: $0° - 32°$. Small for a clinopyroxene.

Dispersion: $r > v$ or $v > r$. Optic axis dispersion may be absent to moderate, and bisectrix dispersion may also be weekly inclined or horizontal.

Twinning: Both simple and multiple twinning on {001} are commonly observed. Exsolution lamellae on {001} may yield the appearance of polysynthetic twinning and, in combination with {100} twinning, produce a herringbone pattern.

Zoning: Zoning is common with Ca poor core and Ca rich rims giving contrasting optical behaviour. Note from the above description that in such a zoned grain (OAP) orientation would be different between core and rim.

Alteration

May alter to fibrous aggregate of amphibole needles or plates (uralite) beginning at the outer margin; less commonly to chlorite or serpentine

Distinguishing Characters

It is identified from other pyroxenes by its small $2V_a < 32°$ and is mainly restricted to volcanic rocks. Orthopyroxene shows lower birefringence and parallel extinction. Olivine is a common associate of pigeonite, showing higher birefringence.

Mineral Association

A characteristic mineral of andesite though not uncommon in basalts. Common associated minerals are augite, calcic plagioclase and orthopyroxenes.

Diopside – Hedenbergite Series

Nature of the Phase

A binary solid solution series. The two end-members are: $CaMg(SiO_3)_2$ (Diopside) and

Ca Fe(SiO$_3$)$_2$ (Hedenbergite). In any natural sample Mn has been observed to be a common content of appreciable amount; Mn-rich variety is called *johannsenite*. Cr^{3+} and Fe^{3+} alongwith Al are common trace elements that influence optical properties of minerals of this series.

Crystallography

Monoclinic; 2/m, C2/c.

Z = 4	Diopside (Di)	Hedenbergite (Hd)
á	1.664	1.732
â	1.672	1.730
ã	1.694	1.755

Plane Polarized Light

Form and Relief: Aggregates of prismatic laths common; cross-sections may be square or octagonal. Relief, moderate to high.

Color and Pleochroism: The color and pleochroism vary with iron content in synthetic samples but color variation, hence pleochroism, is complicated. Diopside is colorless to pale green, with little pleochroism. When pleochroic á = pale to strong green or blue green; â = brownish green or blue green; ã = yellowish green. Noticeable blue and green colors are imparted by substitution of CaCr^{3+}AlSi$_2$O$_6$. It occurs because of a change in coordination and spin state of Cr^{3+} from low spin tetrahedral coordination to high spin octahedral coordination.

Cleavage: Distinct prismatic {110} cleavages intersecting at 93° and 87°. A very prominent basal parting with magnetite or ilmenite dust on parting surface is often present; such grains are called *diallage*.

Under Crossed Polars

Anisotropism: Anisotropic.

Birefringence: Moderate, middle second order green and yellow common. Maximum birefringence may be seen in {010} sections. Change in birefringence has not been found to be governed by ferrous iron substitution, since presence of other ions, especially Al, interfere with optical properties.

δ: 0.031 (Diopside) – 0.024 (Hedenbergite).

Polarization Colors: The order of interference colors changes irregularly from one end of the series to the other. High first order color to low second order commonly seen in prismatic sections but change to first order white in basal sections.

Extinction: Extinction angle is strongly dependent on the section on which it is measured. In {010} section it ranges from 38° – 45°; basal sections {001} show symmetrical extinction to prismatic cleavages, and {100} sections show parallel extinction with prismatic cleavage traces.

Orientation: á ∧ X = 24° – 34°, â = Y, ã ∧ Z = –38° (Diopside) to –48° (Hedenbergite) (Fig. 13.8).

OAP: {010}.

Indicatrix: Biaxial (+). Best view of the optic figure is seen when the section is inclined to {001}; basal section and cleavage fragments yield off-centred figure, and {010} shows flash figure.

Sign of Elongation: Since side-pinacoid, {010}, is the most common habit, and it also shows cleavage traces, it should be the right section to measure elongation sign. However, á and ã being nearly 45° to the cleavage traces, it is not practical to determine so (Fig. 13.9).

Dispersion: $r > v$ weak to strong.

Twinning: Very common lamellar twining on {100} and less common on {001}. Exsolution lamellae of orthopyroxene, and, often of chrome-spinel are present parallel to {010}.

Zoning: Commonly seen though of complex nature because of substitution of several elements.

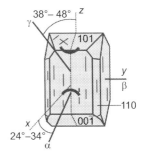

Fig. 13.8 Optic Orientation of diopside-hedenbergite series.

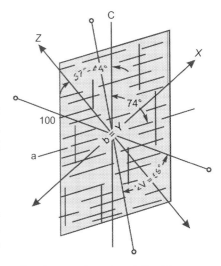

Fig. 13.9 {010} section of diopside-hedenbergite.

Alteration

Common; the product is a mixture of fibrous tremolite, talc, serpentine, or chlorite called *uralite*.

Distinguishing Characters

Can be easily confused with less calcic augite series members. Diopside often occurs in metamorphic rocks while augite is a characteristic mineral of basic igneous rocks.

Inosilicates

Mineral Association

A characteristic calc-silicate mineral occurring with calcite, quartz, forsterite, Ca-garnet, vesuvianite etc.

Augite

Nature of the Phase

Although augite is the most common pyroxene in the terrestrial and lunar basalts, its composition is difficult to define in terms of a single molecular formula because of multi-component solid solution. It is generally classified as Ca-pyroxene but it invariably contains small amount of tetrahedral Al, compensated by the presence of Na in the M2 site. Other common substitutes are Ti and Cr. Hence, the general formula could be [Ca, Na][Mg, Fe^{2+}, Mn, Cr, Ti][Si, Al]$_2$O$_6$.

Crystallography

Monoclinic, 2/m, C2/c.

$$Z = 4$$
$$á = 1.671 - 1.735$$
$$â = 1.672 - 1.741$$
$$ã = 1.703 - 1.774$$

The refractive indices vary as the concentration of various elements varies in one formula unit (Fig. 13.10). It is known that many substituting elements, e.g., Al, Ti, Fe^{3+}, affect indices in opposite direction as substitution in tetrahedral as opposed to octahedral position. Therefore, the utility of refractive index as indicator of composition is limited.

Plane Polarized Light

Form and Relief: Euhedral to subhedral rectangular prismatic laths, cross-sections are square to octagonal (Figs. 13.11A and B). Relief moderate to high commonly varies with Fe content.

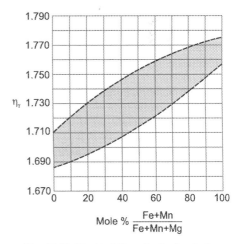

Fig. 13.10 The Variation of refractive indices with compositional changes in augite.

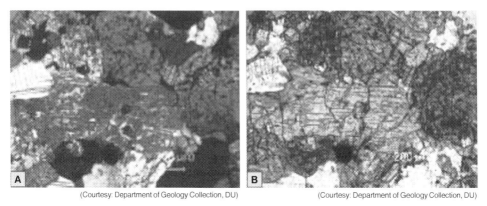

Fig. 13.11 Photomicrographs of an augite grain. **A.** Plane polarized light; **B.** Crossed polars.

Color and Pleochroism: Colorless to grey; non-pleochroic. Ti rich variety, titanaugite is colored and strongly pleochroic; the scheme is á = pale violet-brown, â = violet pink, ã = brown violet. Absorption formula is â > ã > á.

Cleavage: Distinct prismatic cleavages {110}. {010} and {100} are poor cleavage or partings.

Under Crossed Polars

Anisotropism: Anisotropic.

Birefringence: Moderate; maximum birefringence is seen in {010} sections.

ä: 0.018 – 0.030.

Polarization Colors: First order red to low second order color.

Extinction: Maximum extinction angle to {110} cleavage is seen in {010} sections and ranges from 35° – 50°; it is parallel in {100} section and symmetrical in basal sections.

Orientation: á ∧ x = –20° to 35°, â = y, ã ∧ z = –35° to – 50° (Fig. 13.12).

OAP: {010}.

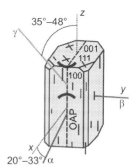

Fig. 13.12 Optic orientation of augite.

Indicatrix: Biaxial (+). Best view of the optic figure is seen when the section is inclined to (100) face. Cleavage planes give off-centred figures. If the microscope stage is rotated till the single isogyre is aligned E-W, note how it is situated. If it coincides with E-W crosswire then the â direction is along the N-S direction and OAP is parallel to {010}.

Inosilicates

$2V_{\tilde{a}}$: 25° but is governed by the Fe^{2+} content of the mineral; can vary up to 61° (Fig. 13.13).

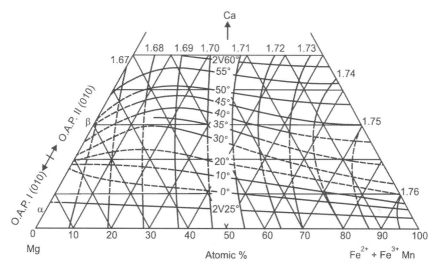

Fig. 13.13 The variation of 2V with Fe^{2+} content in augite. Modified after Hess, H.H., 1949. Amer. Min. 34 621-666.

Sign of Elongation: Because of high angle between cleavage traces in prismatic laths and á or ã directions, sign of elongation is not relevant.

Dispersion: Weak $r > v$, but titanaugite shows strong dispersion $r > v$ with distinct inclined bisectrix dispersion.

Twinning: Simple or repeated twinning on {100} is common. Polysynthetic twinning on {001} combined with twining on {100} produces herringbone structure in {010] sections (*see* previous section of diopside). Interpenetrating twins on {010} and cyclic twinning on {122} are rare.

Zoning: Normal composition zoning with Fe^{2+} increasing towards margin of grains is common. Titanaugite may well show oscillatory or hourglass zoning (Fig. 13.14).

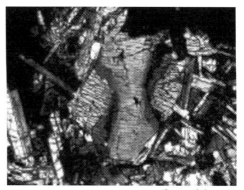

(Courtesy: S. Sengupta)

Fig. 13.14 Photomicrograph of an augite grain exhibiting hourglass zoning under crossed polars.

252
Optical Mineralogy

Alteration

Rims of augite grains often altered and consists of a fibrous aggregate of secondary, light colored amphiboles collectively called uralite; chlorite is a common alteration product of augite; less common products are epidote, carbonates, serpentine, biotite, chalcedony, and talc.

Distinguishing Characters

Augite is easily distinguished from olivine by its lower birefringence and good cleavage. Amphiboles have small extinction angles and cleavage intersections of $56°$ and $124°$.

Mineral Association

Common pyroxene in syenites, gabbro, peridotites and other mafic volcanic rocks. Calcic plagioclase is most common associate along with orthopyroxenes and opaques.

Omphacite

Nature of the Phase

A complex multi-component solid solution. In the simplest term, it may be defined by three-end member compositions, *viz.*, 1. Diopside $Ca(Mg, Fe^{2+})(SiO_3)_2$, 2. Jadeite $(Na\ Al\ (SiO_3)_2$, and 3. Aegirine $Na\ Fe^{3+}\ (SiO_3)_2$.

Crystallography

Monoclinic, 2/m, C2/c or P2/n. The space group depends upon the chemical composition. For the central part of the composition space of the ternary – diopside-jadeite-aegirine system, both octahedral sites, *i.e.*, M_1 and M_2, become ordered, *i.e.*, Ca and Na in M_2 and Mg and Al in M_1. This results in lowering the symmetry from C2/c to P2/n.

$$Z = 4$$
$$á = 1.662 - 1.701$$
$$â = 1.670 - 1.712$$
$$ã = 1.685 - 1.723$$

Plane Polarized Light

Form and Relief: Coarse to medium anhedral grains (Fig. 13.15).

Color and Pleochroism: Colorless to very pale green. In some cases a faint pleochroism has been noticed. $á$ = nearly colorless, $â = ã$ = very pale green.

Cleavage: Typical, distinct, prismatic {110} cleavage.

Under Crossed Polars

Anisotropism: Anisotropic.

Birefringence: Moderate. Maximum birefringence appears in {010} section.

ä: 0.018 – 0.027.

Polarization Colors: First order high to low second order.

Extinction: Symmetrical in basal sections, parallel in {100} sections, and inclined (36° – 45°) in {010} section. Since omphacite is characteristic of high pressure rocks, the grains may develop a series of stacking faults leading to patchy or undulose extinction.

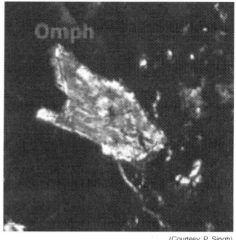

(Courtesy: P. Singh)

Fig. 13.15 Photomicrograph of omphacite in crossed polars

Orientation: á ∧ x = –20° – 32°, â = y, ã ∧ z = –36° – 48° (Fig. 13.16).

OAP: (010).

Indicatrix: The biaxial (+) figure is best seen on (100) and also on (001) in many cases.

2V: 58° – 83°.

Sign of Elongation: May be slighter positive or negative. Usually the slow wave lies near elongation.

Dispersion: Moderate $r > v$; weakly inclined bisectrix.

Twinning: Absent.

Zoning: Not common.

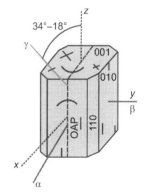

Fig. 13.16 Optic orientation of omphacite.

Alteration

Omphacite is most commonly replaced by secondary, fibrous glaucophane or green barroisite. Alteration is usually takes form of retrograde metamorphic rocks effects and begins at exposed boundaries and fractures.

Distinguishing Characters

Diopside can be confused with omphacite but the latter is usually restricted to eclogites.

Mineral Association

Pyrope garnet and or glaucophane are common associates.

Jadeite

Nature of the Phase

Na Al (SiO$_3$)$_2$. It is almost of stoichiometric composition in most natural occurrences, though it is a multicomponent solid solution with NaAl progressively replaced by CaMg. Fe^{3+}; other metal ions, as well as tetrahedral Al, are also frequently reported.

Crystallography

Monoclinic, 2/m, C2/c.

> Z = 4
> á = 1.640 – 1.681
> â = 1.645 – 1.684
> ã = 1.652 – 1.692

Plane Polarized Light

Form and Relief: Aggregates of anhedral to euhedral grains.

Color and Pleochroism: Most natural examples are colorless under thin section. High ferric content may make jadeite pleochroic.

Cleavage: Typical pyroxene prismatic {110} cleavage.

Under Crossed Polars

Anisotropism: Anisotropic.

Birefringence: Moderate. Best seen in {010} section.

ä: 0.016 – 0.021.

Polarization Colors: Ideal jadeite composition shows maximum birefringence in {010} section with first order yellow. High ferric content causes anomalous blue colors.

Extinction: Maximum extinction angle seen in {010} sections, ranges from 30° – 40°; but is parallel in {100} and symmetrical in basal sections.

Orientation: á ∧ x = –13° to –23°, â = y, ã ∧ z = –30° to –40° (Fig. 13.17).

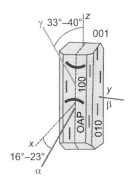

Fig. 13.17 Optic orientation of jadeite.

Inosilicates **255**

OAP: {010}.

Indicatrix: Biaxial positive. Isogyres are broad with only a small number of isochrones. Usually off centred figure seen on basal pinacoid.

Sign of Elongation: Positive (length slow).

Dispersion: $v > r$ moderate; inclined bisectrix dispersion. With increasing Fe^{3+} substitution, dispersion apparently reverses to $r > v$.

Twinning: Fine multiple twinning on {100}, and, rarely on {001} is reported.

Zoning: Not common.

Alteration

Not common, may alter to analcine.

Distinguishing Characters

Cleavage, lower birefringence, and large extinction angle are characteristic.

Mineral Association

A metamorphic mineral representing high pressure-low temperature blueschist facies where it occurs with lawsonite, albite and glaucophane.

Aegirine to Aegirine – Augite

Nature of the Phase

It is essentially a ternary solid solution between $Na\ Fe^{3+}(SiO_3)_2 - (Ca, Na)(Mg,Fe^{2+},Fe^{3+},Al)[(Si,Al)O_3]_2 - NaAl(SiO_3)_2$, *i.e.*, acmite-augite-jadeite molecules.

Crystallography

Monoclinic, 2/m, C2/c.

$Z = 4$		
á	1.750 – 1.766	1.700 – 1.760
â	1.780 – 1.820	1.710 – 1.780
ã	1.795 – 1.836	1.730 – 1.813

The refractive indices can be correlated with Fe^{3+} linearly (Fig. 13.18). However, in low Fe^{3+} bearing varieties, the variation trend becomes uncertain because of other substitution, *e.g.*, $Mg \Leftrightarrow Fe^{2+}$.

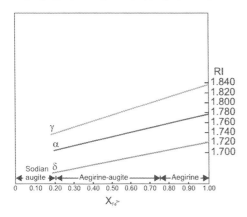

Fig. 13.18 The variation of refractive indices with Fe^{3+} content in aegirine augite.

Plane Polarized Light

Form and Relief: Very similar to eight-sided augite habit, acicular (parallel to z axis) in some cases.

Color and Pleochroism: Bright green color, acmite end-member pale green. á = emerald green, â = Grass green, ã = pale yellow green. Decrease in Fe^{3+} causes change in color and intensity of pleochroism. Absorption scheme á > â > ã.

Cleavage: Prismatic {110} pyroxene cleavage; parting {001} or {100}.

Under Crossed Polars

Anisotropism: Anisotropic.

Birefringence: Strong but is controlled by chemical composition (Fig. 13.19). Maximum birefringence appears in {010} sections.

ä: 0.036 – 0.060.

Polarization Colors: Middle to high second order color. May be masked if original color is highly absorbent.

Extinction: The extinction angle depends upon Fe^{3+} content (Fig. 13.19). For the maximum ferric content it is 10° (á z away from â). As ferric content decreases the extinction angle decrease to zero at Fe^{3+} value 0.7 apfu. With further decrease in ferric content it begins to increase on the opposite side of z to reach a maximum value of 20° at Fe3+ 0.2 apfu. {100} section shows parallel extinction and basal section symmetrical.

Orientation: á ∧ z = –10° to +12°, â = y, ã ∧ x = +28° to +6° (Fig. 13.20). Orientation changes as element concentrations change in the mineral.

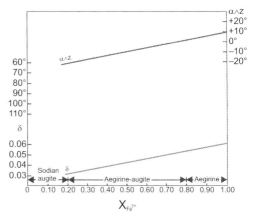

Fig. 13.19 The variation of á ∧ z in aegirine as a function of ferric content.

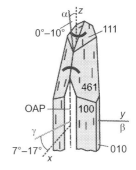

Fig. 13.20 Optic orientation of aeigirine.

Inosilicates

Indicatrix: Biaxial negative for aegirine; aegirine-augite is biaxial positive

2V$_a$: 58° – 90° for aegirine; but for aegirine-augite 2V$_a$ is 90° – 70° (Fig. 13.21). The best figure though off-centred, is seen on basal sections with many colored rings.

Sign of Elongation: Cleavage traces are length fast.

Dispersion: Moderate to strong $r > v$, bisectrix dispersion weekly inclined.

Twinning: Not common, twin plane {100}.

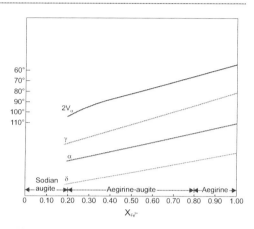

Fig. 13.21 The variation of optic axial angle in aegirine with ferric content.

Zoning: Very common. Fe^{3+} substitution during crystallization of magma causes clinopyroxene to change composition from augite to aegirine. It is obvious in plane polarized light but even more conspicuous in cross polars as different zones have wide range of extinction angle and birefringence.

Alteration

Uralite alteration is very common.

Distinguishing Characters

For a clinopyroxene it has a small extinction angle and biaxial negative character for most of its compositional range. Color and pleochroism along with zoning are the distinguishing characters of aegirine-augite.

Mineral Association

A characteristic mineral of alkali plutons. Riebeckite, nephaline and other Na minerals are common associates.

Wollastonite

Nature of the Phase

A pure phase Ca SiO$_3$.

Crystallography

Triclinic, $\bar{1}$, P$\bar{1}$.

$$Z = 6$$
$$á = 1.616 - 1.640$$
$$â = 1.628 - 1.650$$
$$ã = 1.631 - 1.653$$

Plane Polarized Light

Form and Relief: Columnar bladed to fibrous; grains are elongated along b (Figs. 13.22A and B).

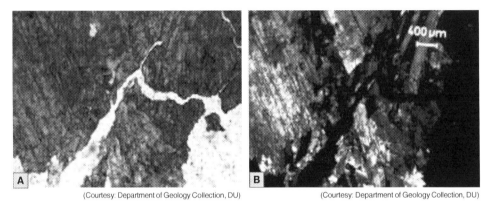

(Courtesy: Department of Geology Collection, DU)   (Courtesy: Department of Geology Collection, DU)

Fig. 13.22 Photomicrographs of wollastonite. **A.** Plane polarized light; **B.** Crossed polars.

Color and Pleochroism: Colorless and non-pleochroic.

Cleavage: Perfect on {100} and good on {001} and {102}; {100} ∧ {001} = 85° and {100} ∧ {102} = 70° in {010} sections.

Under Crossed Polars

Anisotropism: Anisotropic.

Birefringence: Low to moderate; maximum birefringence seen in {010} sections.

ä: 0.013 – 0.014.

Polarization Colors: Low first order colors, may range up to red.

Extinction: Because of more than one set of cleavages in wollastonite, the extinction angles are not easy to define. Each direction of á, â and ã may be nearly parallel to one of the cleavage traces.

Orientation: á ∧ z = 30° – 44°, â ∧ y = 0° – 5°, ã ∧ x = 37° – 50° (Fig. 13.23).

$2V_a$: 35° – 63°.

Inosilicates

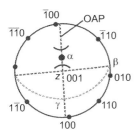

Fig. 13.23 Optic orientation of wollastonite.

Indicatrix: Biaxial negative; best seen on {102} as centred figure with small isogyre separation and few isochrones.

OAP: Nearly parallel to {010}.

Sign of Elongation: Because of large number of cleavage traces it is meaningless to talk about sign of elongation. For while one cleavage will be positively elongated, the other is very nearly of negative elongation.

Dispersion: Weak to moderate $r > v$.

Twinning: Common, twin axis {010} and composition plane {100} multiple twinning.

Zoning: Not common.

Alteration

Not common. In some cases may alter to calcite.

Distinguishing Characters

Habit and cleavages are quite characteristic. Mineral association distinguishes wollastonite from kyanite. Tremolite may confuse an observer but has higher birefringence.

Mineral Association

A typical contact metamorphic or skarn mineral, wollastonite occurs with calcite, tremolite and other calc-silicates.

Sapphirine

Nature of the Phase

A multicomponent solid solution. $(Mg, Fe^{2+}, Fe^{3+}, Al)_8 O_2[(Al, Si)_6 O_{18}]$. Common substitutions are: Mg Fe^{2+}, Al Fe^{3+} with Cr replacing octahedral aluminium. Another important substitution in sapphirine is that of Tschermak, *i.e.*, Mg+Si ⇌ AlAl.

Crystallography

Monoclinic, 2/m, P2$_1$/n or P2$_1$/a.

$$Z = 4$$
$$á = 1.701 - 1.731$$
$$â = 1.703 - 1.741$$
$$ã = 1.705 - 1.745$$

Plane Polarized Light

Form and Relief: Commonly as granular aggregates. Relief moderately distinct.

Color and Pleochroism: Strongly pleochroic from colorless to blue or green. The scheme of pleochroism is a reflection of multicomponent variation in composition: á = colorless, pink, pale yellow. â = sky blue, lavender blue, bluish, bluish green. ã = blue, sapphire-blue, blue green.

Cleavage: {010} moderate; {001}, {100} poor.

Under Crossed Polars

Anisotropism: Anisotropic.

Birefringence: Low; maximum birefringence seen in {010} sections.

ä: 0.005 – 0.009.

Polarization Colors: Low first order grey; in many cases anomalous grey.

Extinction: 6° – 15° with {010} traces.

Orientation: á ∧ x = 41° – 50°, â = y, ã ∧ z = 6° – 9° (Fig. 13.24).

Indicatrix: Biaxial negative to positive depending chemical composition, in particular $X_{Fetotal}$ has significant influence; acute bisectrix figure parallel to {100} with broad isogyres and practically no color rings.

OAP: {010}.

$2V_ä$: For the least $X_{Fetotal}$ $2V_ä$ is about 85° which becomes 45° as $X_{Fetotal}$ increases up to 0.35.

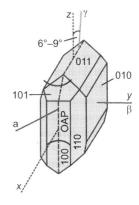

Fig. 13.24 Optic orientation of sapphirine.

Sign of Elongation: Positive (length slow).

Dispersion: Strong v > r; strong optic axis and distinct inclined bisectrix dispersion.

Twinning: Not common. Repeated twins on {100}.

Zoning: Not common.

Alteration

Not common.

Distinguishing Characters

May be confused with corundum which is uniaxial. Kyanite has better cleavages and higher birefringence. Cordierite has practically no relief in comparison with sapphirine.

Inosilicates   **261**

Mineral Association

It is essentially a metamorphic mineral characteristic of granulites that are comparatively Mg-rich. Quartz, sillimanite, cordierite, anthophyllite, spinel etc. are common associates.

Anthophyllite

Nature of the Phase

Most natural samples are close to $\square Mg_7(Si_4O_{11})_2(OH)_2$ composition. Limited solid solution between Mg Fe^{2+} occurs. When X_{Fe} is between 1 to 0.9 prefix ferro- is used; similarly when X_{Fe} is between 0.01 to 0.00 prefix magnesio- is employed. Normally, high Fe content is accompanied by (Mg, Fe^{2+})Si   AlAl substitution (Tschermak substitution) leading to gedrite end member, $\square (Mg, Fe^{2+})_5Al_2[Si_6Al_2O_{22}](OH, F)_2$. When this substitution is not favoured because of external conditions, then, the monoclinic phase grunerite results.

Crystallography

Orthorhombic, 2/m2/m2/m, Pnma.

$$Z = 4$$
$$á = 1.587 - 1.694$$
$$â = 1.602 - 1.710$$
$$ã = 1.613 - 1.722$$

Plane Polarized Light

Form and Relief: A number of habit of anthophyllite are common. Prismatic laths are most common, equally common are fibrous and bladed (Figs. 13.25A and B). Basal section with typical prism outlines terminated by traces of side pincoid may also be seen.

Color and Pleochroism: Normally colorless; substitution of Fe^{2+} may show weak color with a moderate pleochroism in brown. á = pale brown, â = brown, ã = dark brown. Absorption scheme ã = â > á.

Cleavage: {210} perfect (210) $\wedge$ (20) ~ 54.5°.

Under Crossed Polars

Anisotropism: Anisotropic.

Birefringence: Moderate; maximum birefringence seen in {010} sections.

ä: 0.013 – 0.025, decreasing with Fe substitution.

Polarization Colors: Bright second order colors.

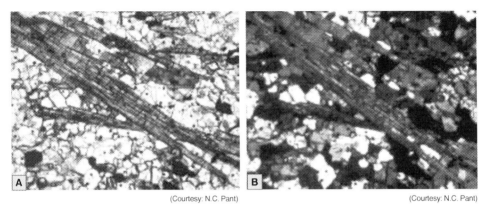

Fig. 13.25 Photomicrographs of anthophyllite. **A.** Plane polarized light; **B.** Crossed polars.

Extinction: Parallel in prismatic sections; symmetrical in basal sections.

Orientation: á = x, â = y, ã = z (Fig. 13.26).

OAP: {010}.

Indicatrix: Biaxial negative for magnesioanthophyllite becoming positive as Fe^{2+} increases. Acute bisectrix figure is best seen in basal sections.

$2V_a$: 115°.

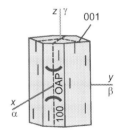

Fig. 13.26 Optic orientation of anthophyllite.

Sign of Elongation: Elongation is positive (length slow).

Dispersion: Weak to moderate $r > v$ or $v > r$.

Twinning: Not present.

Zoning: Not seen.

Alteration

Not common; may alter to talc or serpentine.

Distinguishing Characters

Cleavage and parallel extinction in all [001] sections easily distinguishes anthophyllite.

Mineral Association

A typical high grade mineral in contact and regional metamorphism and skarn deposits. Common associates are cordierite, chlorite, talc, spinel, olivine, orthopyroxenes, plagioclase, quartz.

Cummingtonite-Grunerite

Nature of the Phase

Compositionally related to anthophyllite-gedrite but monoclinic. Since the Mg end member has not been reported in nature, so, it is not exactly a polymorph. $Mg_7(Si_8O_{22})_2(OH)_2 - Fe_7(Si_4O_{11})_2(OH)_2$.

Crystallography

Monoclinic, 2/m, C2/m.

	Cummingtonite	Grunerite
Z = 2		
á	1.632 – 1.663	1.663 – 1.688
â	1.638 – 1.677	1.677 – 1.709
ã	1.655 – 1.697	1.697 – 1.729

Plane Polarized Light

Form and Relief: Columnar, bladed or acicular to fibrous.

Color and Pleochroism: Light colored with weak pleochroism; á = colorless, â = pale yellow-brown, ã = pale green. Absorption formula: ã > â ≥ á.

Cleavage: Distinct prismatic cleavage {110}.

Under Crossed Polars

Anisotropism: Anisotropic.

Birefringence: Moderate to high; best birefringence is seen in {010} sections.

ä: 0.022 (Cummingtonite) – 0.045 (Grunerite).

Polarization Colors: Bright; range from high first order to mid third order.

Extinction: Extinction angle is a function of Fe content; it is 21° in Mg rich to 12° in Fe rich members of the series. Extinction is parallel to cleavages, in {100} sections and symmetrical in basal sections.

Orientation: á ∧ x = –9° to –12°, â = y, ã ∧ z = –21° to 12° (Fig. 13.27).

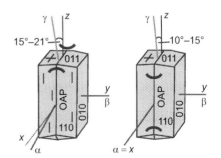

Fig. 13.27 Optic orientation of **A.** Cummingtonite; **B.** Grunerite.

Indicatrix: Biaxial. The optic sign of members of this series changes from positive to negative when grunerite component exceeds 70%.

OAP: {010}.

2V: Cummingtonite has a $2V_{\mathrm{\dot{a}}}$ ranging from $7° – 90°$; grunerite $2V_{\mathrm{\dot{a}}}$ ranges from $90° – 100°$.

Sign of Elongation: Positive (length-slow).

Dispersion: Weak, can be $v > r$ or $r > v$.

Twinning: Simple contact or repeated twining on {100} is very common and twin lamellae are often very narrow.

Zoning: May be zoned by a reaction rim of hornblende, or may form as reaction rims on pyroxenes.

Alteration

Commonly converts to hornblende by reaction rim and may alter to talc or serpentine with iron oxides by-product.

Distinguishing Characters

Inclined extinction distinguishes cummingtonite from anthophyllite. Biaxial positive distinguishes it from tremolite.

Mineral Association

Both end-members have different mineral association though both are essentially metamorphic minerals. Cummingtonite is characteristic of mafic protolith whereas grunerite is a product of metamorphism iron-rich sediments. The former occurs with hornblende, cordierite, garnet, plagioclase etc. while the latter is associated with magnetite, hematite, garnet or even hedenbegite.

Tremolite – Actinolite Series

Nature of the Phase

A part of the calcic amphibole subgroup of the amphiboles. $\square\,Ca_2Mg_5(Si_4O_{11})_2(OH)_2$. In this ideal end-member a number of substitutions occur in natural examples. The common and extensive is Mg Fe^{2+}. Fig. 13.28 gives the scheme of nomenclature for the tremolite-actinolite series.

Inosilicates

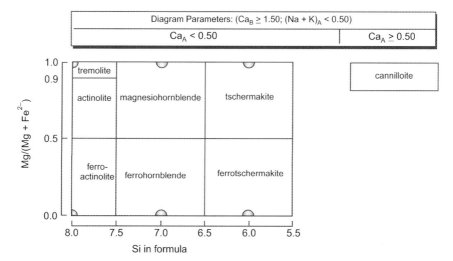

Fig. 13.28 Scheme of Nomenclature of Tremolite-Actinolite Series (After Leake, B.E., Woolley, A.R., Arps, C.E.S., Birch, W.D., Gilbert, M.C., Grice, J.D., Howthorne, F.C., Kato, A., Kisch, H.J., Krivovichev, V.G., Linthout, K. Laird, J. and Mandarino, J., 1997: Nomenclature of amphiboles: Report of the subcommittee on amphiboles of the international mineralogical association commission on new mineral names. Mineralogical Magazine, 61, 295-321).

Crystallography

Monoclinic, 2/m.

$$á = 1.599 - 1.688$$
$$â = 1.610 - 1.697$$
$$ã = 1.620 - 1.705$$

The indices vary systematically as the actinolitic substitution progresses (Fig. 13.29).

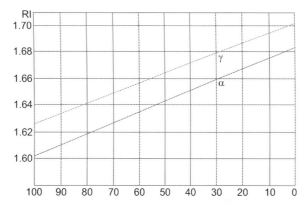

Fig. 13.29 The variation of refractive indices with ferrous iron substitution in tremolite – actinolite series.

Plane Polarized Light

Form and Relief: Prismatic and acicular are common habits. In some cases tremolite also exhibits fibrous habit as well. The relief is moderate to good (Figs. 13.30A, B, C and D).

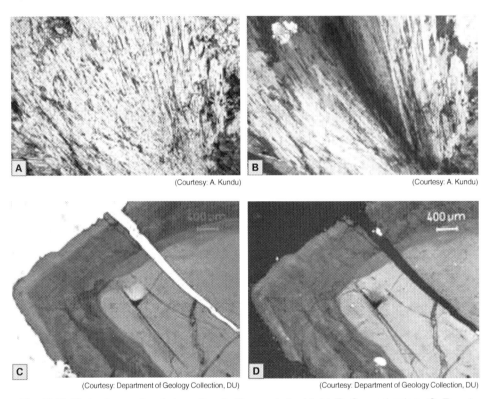

Fig. 13.30 Photomicrographs of tremolite. **A.** Plane polarized light; **B.** Crossed polars. **C.** Zoned actinolite in plane polarized light. **D.** Crossed polars.

Color and Pleochroism: The color of the members of this series is a function of the Fe^{2+}. End-member tremolite is colorless; the color changes to dark shades of green with progressive enrichment of iron. The colored varieties are pleochroic. á = colorless, pale yellow, â = pale yellow-green, ã = green, greenish blue.

Cleavage: Prismatic {110} perfect.

Under Crossed Polars

Anisotropism: Anisotropic.

Birefringence: Moderate. Maximum birefringence seen in {010} sections; decreases with increasing Fe^{2+}.

Inosilicates   267

ä: 0.027 (Tremolite) – 0.017 (Actinolite).

Polarization Colors: Tremolite shows up to middle second order but for actinolite colors may range up to low second order only.

Extinction: Best measured on {010} sections, ranges from about 21° – 13°; the angle is about 15° – 11° on {110}. Increasing actinolitic substitution leads to a small decrease in extinction angle.

Orientation: á $\wedge$ x = –6° to +2°, â = y, ã $\wedge$ z = –21° to – 13° (Fig. 13.31).

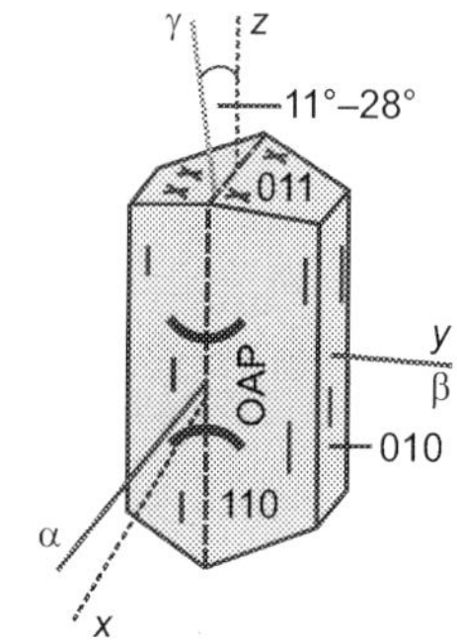

Indicatrix: Biaxial negative; interference figure best seen on (100) face; many colored rings.

$2V_a$: 88° – 75°, lower values are for actinolite.

Fig. 13.31 Optic orientation of tremolite.

Sign of Elongation: Positive (length slow).

Dispersion: Weak, $v > r$, with possible inclined bisectrix effects.

Twinning: Not uncommon on {100}, may be simple or repeated.

Zoning: Pale green actinolite cores may grade to hornblende rims.

Alteration

Tremolite-actinolite may alter to talc, chlorite, or calcite-dolomite.

Distinguishing Characters

Anthophyllite and cummingtonite are biaxial positive, grunerite shows higher birefringence and refractive index.

Mineral Association

A metamorphic mineral characteristic of low grade contact or regional metamorphism. Presence of carbonate minerals in the protolith promotes its growth, hence calcite and talc are common associates. In mafic protoliths, members of this group occur with albitic plagioclase, epidote and chlorite.

Calcic Amphibole ("Common" – Hornblende)

Nature of the Phase

It has already been mentioned above that the present consensus amongst mineralogists favours using *Hornblende* as field name for Ca rich amphiboles with appreciable Si   Al substitution and enough iron to cause dark color. What has been described as hornblende in most literature is a member of at least four component solid solution as shown below:

Magnesiohornblende-Ferrohornblende   □Ca$_2$(Mg, Fe)$_4$Al[Si$_7$AlO$_{22}$](OH)$_2$
Pargasite-Ferropargasite   NaCa$_2$(Mg, Fe)$_4$Al[Si$_6$Al$_2$O$_{22}$](OH)$_2$
Magnesiohastingsite-hastingsite   NaCa$_2$(Mg, Fe)$_4$Fe$_3$+[Si$_6$Al$_2$O$_{22}$](OH)$_2$
Edenite-Ferroedinite   NaCa$_2$(Mg, Fe)$_5$[Si$_7$AlO$_{22}$](OH)$_2$
Tschermakite-Ferrotschermakite   □Ca$_2$(Mg, Fe)$_4$Al$_2$[Si$_6$Al$_2$O$_{22}$](OH)$_2$

A graphical scheme of nomenclature is given in Fig. 13.32.

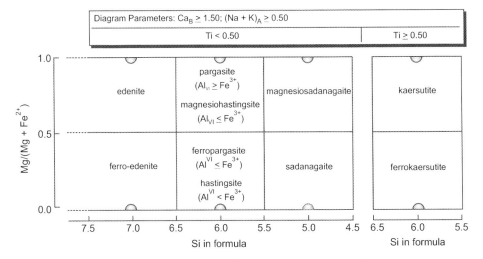

Fig. 13.32 Classification scheme of Calcic Amphiboles (After Leake et al. 1997).

Crystallography

Monoclinic, 2/m, C2/c.

$$Z = 2$$
$$á = 1.61 - 1.728$$
$$â = 1.612 - 1.731$$
$$ã = 1.63 - 1.76$$

Progressive substitution of Fe^{2+} for Mg increases á and ã. Equally effective are tschermak substitution, and Na for Ca in changing values of indices.

Plane Polarized Light

Form and Relief: Prismatic laths are common along with basal cross-section (Figs. 13.33A, B and C).

Inosilicates

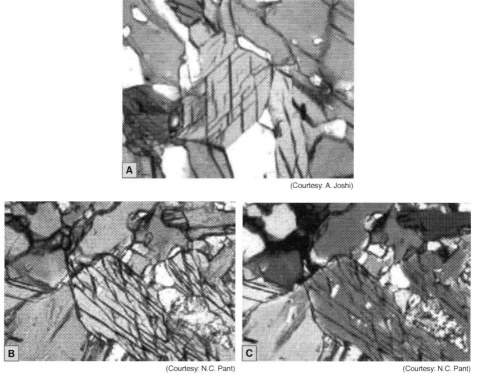

Fig. 13.33 Photomicrographs of calcic amphibole grains. **A.** Plane polarized light; **B.** Plane polarized light; **C.** Crossed polars.

Color and Pleochroism: Highly variable color in green or brown revealing the control of chemical composition on color of the mineral. Fe^{2+}, Fe^{3+} and Ti amounts per formula unit greatly affect the color and its intensity. Strongest absorption in '*hornblende*' is either along ã or â. Strongly pleochroic; pleochroic schemes of many representative natural '*hornblende*' are given in the Table 13.1.

Oxyhornblendes are highly absorbent from brown to black; á = yellow, â = chestnut brown, ã = darkbrown or black.

Cleavage: Prismatic cleavages {110}; paratings along {100} and {001}.

Under Crossed Polars

Anisotropism: Anisotropic.

Birefringence: Moderate; best seen on (010) sections.

ä: 0.015 – 0.034.

Polarization Colors: Range from high first order to low second order.

Table 13.1 Some examples of pleochroism and absorption in calcic amphiboles.

Mineral	Pleochroic Scheme			Absorption formula
	á	â	ã	
Magnesiohornblende-Ferrohornblende	Green, Pale green,	Dark green, green	Black, yellow brown-brown	ã ≥ â > á or â > ã > á
Pargasite-Ferropargasite	Colourless, Greenish yellow,	Light brown, bluish green	Light brown, bluish green	â = ã > á
Magnesiohastingsite Hastingsite	Yellow, greenish brown	Deep greenish blue, greenish green	Deep olive-green smoky blue green	ã > â > á

Extinction: 12° – 34° in {010} drops to 0° in {100} sections, and symmetrical in basal sections.

Orientation: á ∧ x = +3° to –19°, â = y, ã ∧ z = –12° and –34° (Figs. 13.34A, B and C).

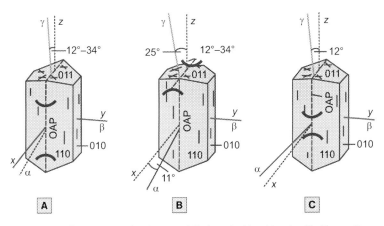

Fig. 13.34 Orientation of calcic amphiboles. **A.** Hornblende; **B.** Pargasite; **C.** Hastingsite.

OAP: {010}.

Indicatrix: Biaxial negative; Mg-rich pargasite is biaxial positive.

Inosilicates   271

2V: It is dependant on the chemical composition of the calc-amphibole species. For most calcic amphiboles $2V_a$ ranges from $90° - 44°$; however, as mentioned above, Mg-rich pargasite is optically positive, hence, its $2V_a$ ranges $56° - 90°$; hastingsite shows $2V_a$ form $30° - 0°$.

Sign of Elongation: Positive (length slow) for prismatic section.

Dispersion: $v > r$ or, more rarely $r > v$, and a few varieties, i.e., edenite, ferrohastingite show pronounced inclined bisectrix dispersion.

Twinning: Not commonly.

Zoning: Color zoning in green or brown to green is likely with dark zones at either core or margin.

Alteration

Normally alters to chlorite with associated calcite and epidote.

Distinguishing Characters

Intense color and high refraction index distinguishes hornblende from iron free amphiboles.

Mineral Association

A very common mineral in the crust of the earth. It occurs in acid igneous rocks like granite where the associated minerals are quartz, plagioclase, K-feldspar. It is the primary mineral of a metamorphic rock called amphibolite where it occurs with plagioclase, and garnet.

Glaucophane

Nature of the Phase

A sodium amphibole; the end-member composition is $\square Na_2 Mg_3 Al_2(Si_4O_{11})_2 (OH)_2$; generally it forms major solid solutions with magnesio-riebeckite ($\square Na_2 Mg_3Fe_2^{3+}(Si_4O_{11})_2(OH)_2$) and riebeckite ($\square Na_2Fe_3^{2+} Fe_2^{3+}(Si_4O_{11})_2 (OH)_2$); intermediate compositions are known as crosstie (Fig. 13.35). Here, we shall describe these two Na-amphiboles separately because of distinctly different geologic mode of occurrence.

Crystallography

Monoclinic, 2/m, C2/m.

$$Z = 2$$
$$á = 1.594 - 1.630$$
$$â = 1.612 - 1.650$$
$$ã = 1.618 - 1.652$$

Refractive indices vary with Mg Fe^{2+} and Al Fe^{3+} substitution (Fig. 13.36).

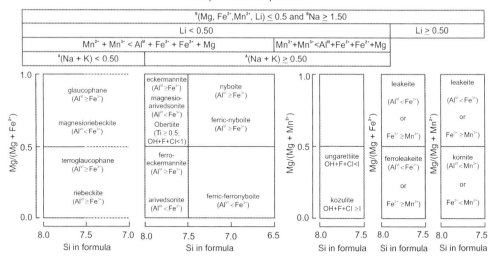

Fig. 13.35 Classification Scheme of Sodic Amphiboles (After Leake, B.E., Woolley, A.R., Birch, W.D., Burke, E.A.J., Ferraris, G., Grice, J.D., Hawthorne, F.C., Kisch, H.J., Krivovichev, V.G., Sachumacher, J.C., Stephenson, N.C.N. and Whittaker, E.J.W., 2004: Nomenclature of amphiboles: addition and revision to the International Mineralogical Association amphibole nomenclature. Mineralogical Magazine, 68, 209-215).

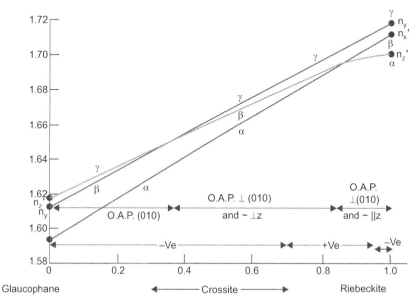

Fig. 13.36 Variation in the refractive indices of glaucophane-riebeckite minerals (Modified after Borg, I.Y., 1967, Contr. Min. Petr., 15, 67-92).

Plane Polarized Light

Form and Relief: Acicular to prismatic laths, also basal sections (Fig. 13.37).

Color and Pleochroism: Either blue or bluish green with very strong pleochroism. á = colorless, pale green, â = violet, ã = blue. Absorption Formula: ã > â > á

Cleavage: Typical prismatic {110} cleavage.

(Courtesy: P. Singh)

Fig. 13.37 Photomicrograph of glaucophane in crossed polars.

Under Crossed Polars

Anisotropism: Anisotropic.

Birefringence: Moderate, best seen on {010} sections. Mg rich varieties show greater birefringence.

Polarization Colors: May be masked by the intense deep blue mineral color.

Extinction: Commonly 10° or less depending upon the composition.

Orientation: á ∧ x = +8° – 5°, â = y, ã ∧ z = –6° – 9° (Fig. 13.38).

OAP: {010}.

Indicatrix: Biaxial negative, centred figure when obtained in {100} section shows a few colored rings; generally intense color masks the figure.

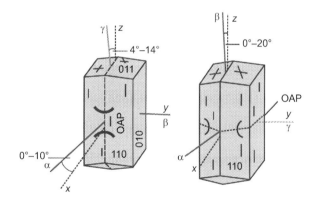

Fig. 13.38 Optic orientation of glaucophane and crossite.

$2V_{\acute{a}}$: 50° – 0°. The optic axial angle of sodic amphibole is very sensitive to Al: Fe^{3+} ratio, so much so that á changes from being acute bisectrix to obtuse and then back to acute bisectrix near end-member riebeckite.

Sign of Elongation: Positive; cleavage traces on {010} are length slow.

Dispersion: Weak, $v > r$.

Twinning: Simple as well as lamellar; twin plane {100}.

Zoning: Composition zoning is common and appears as concentric layers of other amphiboles particularly calcic varieties (Fig. 13.39).

Alteration

Glaucophane may be replaced by green amphibole, presumably actinolite.

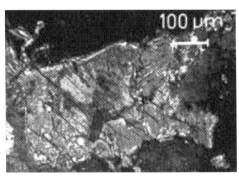

(Courtesy: P. Singh)

Fig.13.39 Photomicrograph of a zoned glaucophane in crossed polars.

Distinguishing Characters

Prismatic habit distinguishes glaucophane from riebeckite.

Mineral Association

A characteristic mineral of blueschist facies of metamorphism. Common associates are epidote, lawsonite, jadeite etc.

Riebeckite

Nature of the Phase

Riebeckite ($\square$ Na$_2$Fe$_3^{2+}$ Fe$_2^{3+}$ (Si$_4$O$_{11}$)$_2$(OH)$_2$. Forms solid solution with other sodium amphiboles. The variation in chemical composition in riebeckite, as well as its effect on optical properties has already been described under glaucophane.

Crystallography

Monoclinic, 2/m, C2/m.

$$Z = 2$$
$$á = 1.690 - 1.702$$
$$â = 1.690 - 1.712$$
$$ã = 1.702 - 1.719$$

Plane Polarized Light

Form and Relief: Acicular to fibrous. Relief moderate to high.

Inosilicates 275

Color and Pleochroism: Intense blue. Highly pleochroic. á = dark blue, â = dark grey blue, ã = bluish brown. ã > â > á.

Cleavage: On prismatic laths one may see prismatic {110} cleavages.

Under Crossed Polars

Anisotropism: Anisotropic.

Birefringence: Moderate, seen on {010} section. Mg-rich varieties showing greater birefringence.

ä: 0.012 – 0.017.

Polarization Colors: High first order to low second order, but often well obscured by deep blue mineral color.

Extinction: Extinction angle of $7° - 8°$ in {010} sections.

Orientation: á $\wedge$ z = $-8°$ to $7°$, â = y, ã $\wedge$ x = $+6°$ to $+7°$ (Fig. 13.40). Optic figure not easily seen because of intense deep blue color.

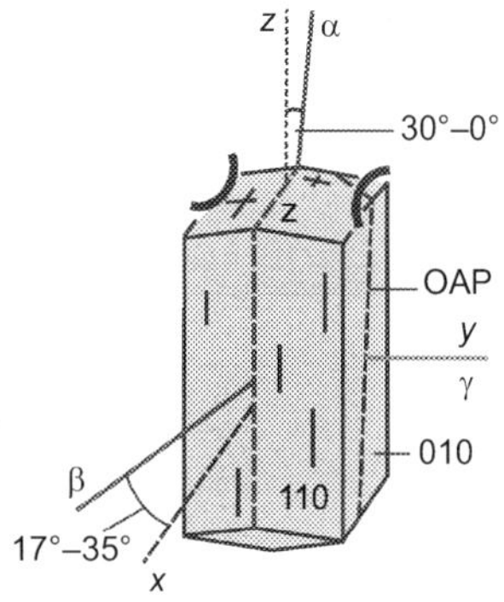

Indicatrix: Biaxial positive for the Mg rich varieties becoming negative towards the end member composition.

Fig. 13.40 Optic orientation of riebeckite.

OAP: Normal to {010}.

$2V_{\text{á}}$: $0° - 90°$.

Sign of Elongation: Negative elongation (length fast)

Dispersion: Strong dispersion $r > v$ and $v > r$.

Twinning: Simple and lamellar on {100}.

Zoning: Composition zoning is common and appears as concentric layers or irregular patches of different color extinction position.

Alteration

Riebeckite commonly alters to ferric oxides or siderite with release of free silica. The fibrous silica "tiger eyes" result from the alteration of fibrous riebeckite

Distinguishing Characters

Habit and color easily distinguish riebeckite.

Mineral Association

A common mineral of alkali igneous rocks. Common associate are aegirine-augite, plagioclase and quartz in most cases.

■■■

14

Phyllosilicates

- Serpentine
- Talc
- Muscovite
- Paragonite
- Biotite
- Chlorite
- Gluconite
- Apophyllite
- Stilpnomelane
- Prehnite

(Courtesy: D.M. Bannerjee)

A photomicrograph of a sandstone with abundant detrital gluconite grains. (Plain polarized light).

Technical Text Box 14.1

The basic unit of silicates, silicon tetrahedron, is linked to other tetrahedra by sharing all three basal oxygens in such a cyclic manner that these form a layer. The tetrahedral layer is linked to an octahedral layer which may consist of tri- or di- valent metal ions linked to six oxygens. Some of this oxygen may be shared with the tetrahedra. This combination is known as t-o sheet. This may be linked to other layers along z-direction. The manners of stacking of these layers are numerous giving rise to a large number of phyllosilicate species.

The sheet structure creates greater polarization of atoms as compared to planes normal to it. This factor retards the vibration of light in x-y plane; hence, in general, phyllosilicates are optically negative with the exception of apophyllites and Fe-rich chlorites.

The well-known tetrahedral-octahedral layers mismatch sets up internal stresses that retard the growth of crystals of large grain size. Therefore, phyllosilicates are of small grain size with majority of species falling within the clay fraction range. Muscovite is the notable exception.

© The Author(s) 2023
P. K. Verma, *Optical Mineralogy*,
https://doi.org/10.1007/978-3-031-40765-9_14

Serpentine

Nature of the Phase

$Mg_3Si_2O_5(OH)_4$. Small variation in composition has been noticed; commonly a pure phase. Three varieties of serpentine, *viz.*, antigorite, lizardite and chrysotile are distinguished by physical deformation, disposition of the basic two layer, 7 Å sheet structure and concentration ranges in certain chemical components.

Crystallography

The basic two-layer unit is monoclinic, but polytypes may show different symmetry from trigonal to monoclinic or orthorhombic.

	antigorite	lizardite	chrysotile
Z	2	2	4
á	1.558 – 1.567	1.538 – 1.554	1.532 – 1.545
â	~1.560	—	—
ã	1.562 – 1.574	1.546 – 1.560	1.545 – 1.556

The determination of refractive index, in fact, that of most other optical parameters, is rendered difficult, because of the fine-grained nature of serpentine minerals. Antigorite is the coarsest of all these minerals, hence, its low relief can get noticed if carefully observed. However, as the above data indicate, antigorite and lizardite indices are close, making it irrelevant as a distinguishing property.

Plane Polarized Light

Form and Relief: Chrysotile is commonly asbestiform, as parallel fibres (Figs. 14.1A and B) as aggregate with cross fibres, commonly as distinct veins; the fibres show

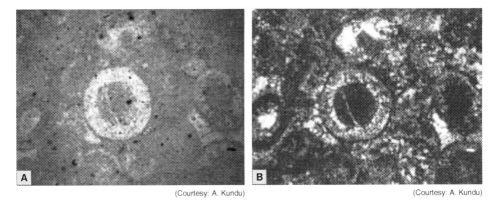

(Courtesy: A. Kundu)   (Courtesy: A. Kundu)

Fig. 14.1 Photomicrographs of a section cut normal to chrysotile fibres. **A.** Plane polarized light; **B.** Crossed polars.

different relief parallel and normal to fibre axis. Antigorite is foliated as large sheet or as scaly or shredded aggregates. Lizardite is largely massive with netlike mesh structure and undulatory extinction. Al-lizardite forms large 'books'.

Color and Pleochroism: Serpentine is colorless to pale green, with weak pleochroism. ã > â > á. á = colorless to pale yellow-green, â = colorless to pale yellow green, and ã = yellow to light green.

Cleavage: Perfect basal {001} for antigorite and lizardites. Fibres of chrysotile are parallel to x-direction.

Under Crossed Polars

Anisotropism: Anisotropic.

Birefringence: Low; basal flakes are practically isotropic.

ä: Antigorite = 0.004 – 0.000; Lizardite = 0.006 – 0.008; Chrysotile = 0.013 – 0.017.

Polarization Colors: Gray to white of first order and may be anmolous, as pale yellow or pale olive.

Extinction: Undulose extinction is the rule in various serpentine minerals; it arises because of internal deformation of sheets constituting the mineral.

Orientation: á ∧ z = 0° to 7°, â = y, ã ≈ x (Fig. 14.2).

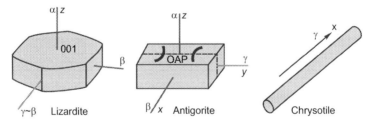

Fig. 14.2 Optic orientation of serpentine minerals.

OAP: In antigorite it is normal to (010) plane.

Indicatrix: Biaxial negative. Interference figure commonly not observed because of fine grained size.

2V$_á$: Highly variable, may even be zero in some cases but values between 20° and 80° are most common.

Sign of Elongation: Chrysotile fibres are length fast.

Dispersion: Very weak. r > v.

Optical Mineralogy

Twinning: Sub-microscopic twinning, repeated twinning is probably very common and may be responsible for antigorite sheets.

Zoning: Not seen.

Alteration

Not common.

Distinguishing Characters

Chrysotile fibres may be confused with asbestiform amphiboles, but the latter minerals show inclined extinction, higher refractive index, and higher birefringence. The best method to distinguish these minerals is either x-ray microbeam method or scanning electron microscopy. Optical methods are not conclusive owing to complications in the stacking of sheets in the unit cell, *e.g.*, normally lizardite and antigorite are length slow, but in several instances stacking of platelets normal to fibre axis may yield length fast character in lizardite.

Mineral Association

Relict olivine and pyroxenes are common; calcite and other carbonates are often found with serpentine minerals.

Talc

Nature of the Phase

Normally a pure phase: $Mg_3Si_4O_{11}(OH)_2$.

Crystallography

Triclinic or monoclinic, $C\bar{1}$, C2/c or Cc.

$$Z = 1 \text{ (Triclinic)}; 2 \text{ (Monoclinic)}$$
$$á = 1.539 - 1.550$$
$$â = 1.589 - 1.594$$
$$ã = 1.589 - 1.600$$

Plane Polarized Light

Form and Relief: Aggregates of tiny foliated flakes. Relief low to negative Figs. 14.3A and B).

Color and Pleochroism: Colorless and non-pleochroic.

Cleavage: Perfect basal cleavage {001}.

Phyllosilicates

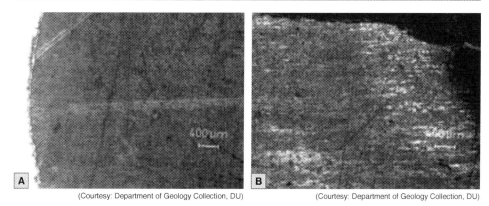

(Courtesy: Department of Geology Collection, DU)   (Courtesy: Department of Geology Collection, DU)

Fig. 14.3 Photomicrographs of talc. **A.** Plane polarized light; **B.** Crossed polars.

Under Crossed Polars

Anisotropism: Anisotropic.

Birefringence: Strong. Usually only basal sections are available owing to excellent cleavage on which very low birefringence (0.0006 or less) is observed. Maximum birefringence, seen in {010} sections or essentially any section normal to cleavage, is strong.

Polarization Colors: Low third order colors.

ä: ~ 0.05.

Extinction: Extinction is essentially parallel to cleavages and

Orientation: á $\wedge$ z = –10°, â = x, ã ≈ y (Fig. 14.4).

OAP: It is normal to {010} plane.

Indicatrix: Biaxial negative. Large basal section would reveal the well centred figure with a few isochrones.

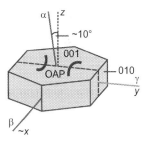

Fig. 14.4 Orientation of talc.

$2V_{\acute{a}}$: 0° – 30°. Low $2V_{\hat{a}}$ may result from offsets in stacking of layers.

Sign of Elongation: Elongation is positive (length slow).

Dispersion: Strong dispersion, $r > v$.

Twinning: Not present.

Zoning: Not seen.

Alteration

Not present.

Distinguishing Characters

Muscovite and pyrophyllite have larger $2V_a$ whereas phogophite and phengites show higher refractive indices. Fine grained talc and sericite have close optical properties though commonly mineral associations are different. Hence, a check from x-ray diffraction pattern is useful.

Mineral Association

A common metamorphic and metasomatic mineral. Association with carbonates, tremolite and olivine (often relict) is characteristic.

Muscovite

Nature of the Phase

Commonly occurs as pure phase or nearly so. $KAl_2(Si_3Al)O_{10}(OH)_2$ Sodium analogue, paragonite, also is fairly common but as a distinct species. There is a large miscibility gap between the two minerals at room temperature and pressure.

Crystallography

Monoclinic, 2/m, C2/m.

$$Z = 1$$
$$á = 1.552 - 1.576$$
$$â = 1.582 - 1.615$$
$$ã = 1.587 - 1.618$$

The refractive indices are functions of substitution of several elements in the structure of muscovite. For example, increasing amount of Fe^{3+} and Mn, and decreasing amount of Al increases the refractive indices.

Plane Polarized Light

Form and Relief: Coarse, micaceous, with one set of cleavage planes often bent by compression, mica 'fish' form is also a common habit; flakes are often oriented forming prominent lineation. Fig 14.5 illustrates some of the many habits of this mineral. Muscovite may occur as fine scales coating on grains or as inclusions, this variety is named sericite, though strictly speaking, it is a different mineral with its own composition. Some secondary muscovite, usually appears as fibrous aggregates.

Phyllosilicates

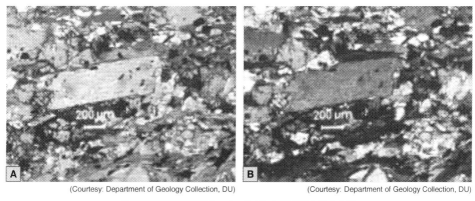

Fig. 14.5 Photomicrographs of muscovite. **A.** Plane polarized light; **B.** Crossed polars; **C.** Muscovite 'fishes' in a schist.

Color and Pleochroism: Colorless and non-pleochroic. However, traces of chromium imparts green color to muscovite; the variety is known as fuchsite. Its pleochroism is given: á = colorless to light green, â = green, ã = dark green.

Cleavage: Perfect basal {001} cleavage.

Under Crossed Polars

Anisotropism: Anisotropic.

Birefringence: Maximum birefringence seen in {100} sections is moderate to strong; all sections normal to cleavage show nearly maximum birefringence.

ä: $0.036 - 0.049$.

Polarization Colors: High second and third order colors.

Extinction: Close to parallel extinction. Maximum extinction angle with cleavage traces is $0° - 3°$, seen on {010} sections. Bent and wrinkled sheets cause wavy extinction.

Orientation: á ∧ z = 0° to –5°, â ∧ x = +1° to +3°, ã = y (Fig. 14.6).

OAP: It is normal to {010} plane.

Indicatrix: Biaxial negative. Excellent centred interference figure seen normal to basal section with numerous colored rings; melatopes are well separated. Some polytypes show uniaxial figures; phengite characteristically exhibit smaller 2V_a.

2V_a: 30° – 47°.

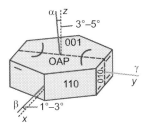

Fig. 14.6 Optic orientation of muscovite.

Sign of Elongation: Positive (length-slow) to cleavages traces.

Dispersion: Optic bisectrix dispersion is rather distinct r > v.

Twinning: The complex mica twin is formed about a twin axis lying in the {001} composition plane Normal.

Zoning: Not present.

Alteration

Not common.

Distinguishing Characters

Large grain size as compared to other phyllosilicates such as talc, vermiculite, colorless phlogopite and clintonite as well as larger 2V_a distinguish muscovite. Paragonite is optically indistinguishable from muscovite.

Mineral Association

Wide range of minerals may occur with muscovite, though essentially a high Al composition favours muscovite; therefore carbonate rocks and their metamorphic products normally do not carry muscovite.

Paragonite

Nature of the Phase

Commonly occurs as pure phase or nearly so. $NaAl_2(Si_3Al)O_{10}(OH)_2$

Crystallography

Monoclinic, 2/m, C2/c.

Phyllosilicates

> Z = 1 (2M₁ polytype)
> á = 1.566 – 1.580
> â = 1.594 – 1.609
> ã = 1.600 – 1.609

Plane Polarized Light

Form and Relief: Fine-grained with low relief.

Color and Pleochroism: Colorless and non-pleochroic.

Cleavage: Perfect basal {001}.

Under Crossed Polars

Anisotropism: Anisotropic.

Birefringence: Maximum birefringence seen in {100} sections is moderate; all sections normal to cleavage show nearly maximum birefringence.

Polarization Colors: High second and third order colors.

ä: 0.028 – 0.038.

Extinction: Close to parallel extinction. Maximum extinction angle with cleavage traces is 0° – 5°, seen on {010} sections.

Orientation: á ∧ z = 0° to –5°, â ∧ x = ~ 0°, ã = y (Fig. 14.7).

OAP: It is normal to {010} plane.

Indicatrix: Biaxial negative. Excellent centred interference figure seen normal to basal section with numerous colored rings. Some polytypes show uniaxial figures; phengite characteristically exhibits smaller 2V_á.

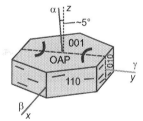

Fig. 14.7 Optic orientation of paragonite.

2V_á: 0° – 40°.

Sign of Elongation: Positive (length-slow) to cleavages traces.

Dispersion: Optic axis dispersion is rather distinct $r > v$.

Twinning: The complex mica twin is formed about a twin axis lying in the {001} composition plane.

Zoning: Not present.

286 — Optical Mineralogy

Alteration

Not common.

Distinguishing Characters

Paragonite is optically indistinguishable from muscovite, 2V is a distinguishing character but paragonite being fine-grained, the determination becomes very difficult. EPMA and x-ray diffraction are the best methods for distinguishing paragonite from muscovite.

Mineral Association

Muscovite, and phengite are common associates.

Biotite

Nature of the Phase

A four-component solid solution. The end-members are:

Phlogopite	:	$K_2Mg_6(Si_6Al_2)O_{20}(OH, F)_4$
Annite	:	$K_2Fe_6^{2+}(Si_6Al_2)O_{20}(OH, F)_4$
Eastonite[1]	:	$K_2(Mg_5Al)(Si_5Al_3)O_{20}(OH, F)_4$
Siderophyllite	:	$K_2(Fe_5Al)(Si_5Al_3)O_{20}(OH, F)_4$

Crystallography

Monoclinic, 2/m, C2/m.

$$Z = 1$$
$$á = 1.530 - 1.625$$
$$â = 1.557 - 1.696$$
$$ã = 1.558 - 1.696$$

To a certain extent composition controls the value of refractive index. Fe^{2+} substitution for Mg increases the index, which also increases as Mn, Ti and Fe^{3+} also substitute for Mg; F substitution decreases the index. It is recommended that x ray diffraction studies are more reliable for X_{Fe} determination than RI of a biotite sample.

---

1. Eastonite is no longer a valid mineral name. The natural biotite in majority of the cases, has the composition which is away from three of these four corners, therefore, the biotite solid solution is largely restricted to phlogopite-annite binary join; even in this binary, natural biotite samples do not reach up to annite end.

Plane Polarized Light

Form and Relief: Fine to coarse-grained thin to broad flakes or laths; subhedral to euhedral clusters of biotite laths having non-parallel cleavages in adjacent laths are common; biotite also occurs as 'fishes'. Cleavage fragments also occur though not so common. Inclusions of radioactive element bearing minerals, particularly, zircon, commonly occur; particle emissions from these inclusions burn out the adjacent parts of the host mineral creating a pleochroic halo. Fig. 14.8 illustrates some of these habits. Relief is moderate.

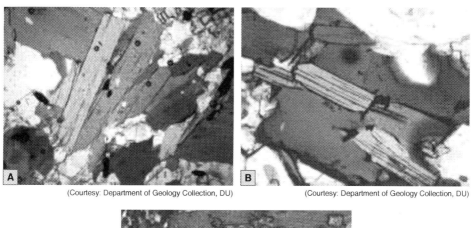

(Courtesy: Department of Geology Collection, DU)   (Courtesy: Department of Geology Collection, DU)

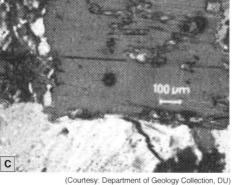

(Courtesy: Department of Geology Collection, DU)

Fig. 14.8 Photomicrographs of biotite. **A.** Plane polarized light; **B.** Crossed polars; **C.** With pleochroic halos in plane polarized light.

Color and Pleochroism: Highly colored; brown and green colors are most common with former predominating over the latter. Influence of composition and in particular of oxidation state of transition metal ions is quite obvious. Fe^{3+} causes olive green color or even brown, Ti makes it reddish brown. When Fe^{3+} is absent but Mn^{2+} and Ti^{2+} are present the color is green and it becomes progressively darker with the incorporation

Fe^{2+}. Large amount of these metals make it highly absorbent. When it is not strongly absorbent it is strongly pleochroic in shades of brown or green. Absorption Formula is ã ≅ â>>á. Pleochroic scheme is á = pale green, light brown; â = blue green, brown; ã = dark green, dark brown. Fe^{3+} and Fe^{2+} are often present on adjacent octahedral sites, and when these exchange electrons, a significant absorption of incident light beam would occur making such grains highly absorbent.

Cleavage: Perfect basal {001} cleavage.

Under Crossed Polars

Anisotropism: Anisotropic.

Birefringence: Highly variable depending upon orientation of the section being viewed. Being a pseudohexagonal mineral, the basal sections show practically nil birefringence. Sections normal to cleavage show maximum birefringence. Very high birefringence.

ä: 0.04 – 0.07.

Polarization Colors: Bright, third order colors. Polarization colors may be masked by the high absorbent nature of grains.

Extinction: Maximum extinction angle is 0° – 9° from cleavage in {010} sections. Wavy extinction results from bent cleavages.

Orientation: á ∧ z, â = y, ã ∧ x = 0° to –9° (Fig. 14.9).

OAP: {010}. Most of the biotite are 1M polytypes with the optic plane parallel to {010}, but few are 2M polytypes with optic plane normal to {010}, and rare trigonal forms have no optic plane.

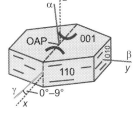

Fig. 14.9 Optic orientation of biotite.

Indicatrix: Biaxial negative. Because of the small $2V_a$ values it may yield nearly uniaxial figure; a careful rotation of the microscope stage would reveal a small isogyre separation. It is difficult to obtain a figure when the grains are highly absorbent.

$2V_a$: 0° – 25°.

Sign of Elongation: Positive (length slow) for sections cut normal to {001} cleavage.

Dispersion: Optic axis dispersion is generally weak and usually v > r, however, some Mg rich biotites show r > v.

Twinning: Not common.

Zoning: Not common.

Phyllosilicates 289

Alteration

Not very common in igneous biotites. Detrital biotite in sediments show biotite altering to hydrobiotite and then to vermiculite. In metamorphic rocks, biotite may be altered hydrothermally to chlorite via green biotite with magnetite and rutile as by product.

Distinguishing Characters

Identified by greater absorption, darker color and higher refractive index; pleochroic halos around inclusions of zircon and other radioactive element minerals (Fig. 14.8C) are also characteristic of biotite.

Mineral Association

Biotite is a mineral of wide occurrence because of its large compositional range. It occurs in quartz rich igneous rocks as well as in ultrabasic peridotites. It is a very common constituent of metapelites where it occurs with quartz, plagioclase, epidote, garnet, staurolite, sillimanite, kyanite, etc. It is also a common mineral of amphibolites.

Chlorite

Nature of the Phase

In fact, chlorite is a group name for a large number of minerals which are sheet silicates. The least complicated chlorite species consists of a combination of two sheets: one is a t-o-t ('talc') sheet, and another an octahedral ('brucite') sheet. Extensive substitution in octahedral and tetrahedral sites is a rule rather than an exception. The led to proliferation of names. The general formula may be written as: $(Mg, Fe^{3+}, Fe^{2+}, Mn, Al)_{12}[(Si, Al)_8 O_{20}](OH)_{16}$.

Crystallography

Monoclinic, 2/m, C2/m, also as $C\bar{1}$ or C1 in some cases.

$$Z = 1$$
$$á = 1.57 - 1.67$$
$$â = 1.57 - 1.69$$
$$ã = 1.57 - 1.69$$

With progressive substitution of Fe for Mg, and Al for Si there is systematic increase in refractive index (Fig. 14.10). When number of cations for Fe and Mg are nearly equal, basal sections may appear isotropic owing to close values of â and ã.

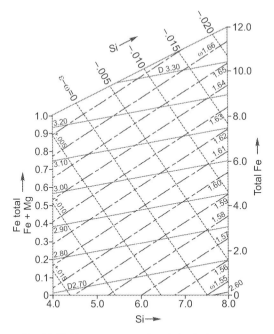

Fig. 14.10 Variation in optical and physical properties in relation to composition in chlorites. (After Hess, 1954).

Plane Polarized Light

Form and Relief: Most common habit is shredded aggregates of fine grained scales. Also as broad laths (Fig. 14.11). Often occurs as pseudomorph after other minerals like biotite, garnet, cordierite etc.

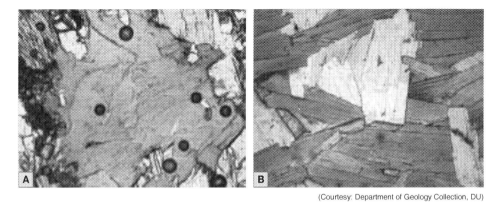

(Courtesy: Department of Geology Collection, DU)

Fig. 14.11 Photomicrographs of chlorites. **A.** Plane polarized light; **B.** Crossed polars.

Color and Pleochroism: Color is controlled by its composition though the variation is within the shades of green. Fe^{2+}, Mn and Ti^{4+} substitution causes a darker color and slight pleochroism. á = â > ã. á colorless, pale green = â, ã = colorless, pale yellow.

Cleavage: Perfect basal cleavages {001}.

Under Crossed Polars

Anisotropism: Anisotropic.

Birefringence: Low. Best seen in sections normal to cleavage.

ä: 0.00 – 0.02.

Polarization Colors: Low first order, rarely exceeding yellow. Anomalous in many cases.

Extinction: 0° – 7° normal to cleavages. Multiple twinning may cause wavy extinction.

Orientation: á ∧ x = 7° to –2°, â = y, ã ∧ z = 0° to –9° (positive varieties). á ≈ z, â = y, ã ∧ x = 0° to –7° (positive varieties). See Fig. 14.12 for orientation. Iron rich chlorites are optically negative in contrast to magnesium rich chlorite that are optically positive.

OAP: {010} or almost parallel to it.

Indicatrix: Biaxial positive as well as negative. Optic figure best seen in basal sections.

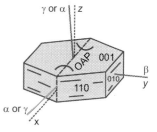

Fig. 14.12 Optic orientation of chlorites.

Sign of Elongation: Negative elongation (length fast) for biaxial positive chlorites and positive elongation (lengths slow) for biaxial negative chlorites.

Dispersion: Moderate dispersion $v > r$.

Twinning: Common, multiple on {001}, also on other laws, e.g., [310] twin axis with {001} as composition plane.

Zoning: Not seen.

Alteration

Not common.

Distinguishing Characters

Green color, low birefringence and anomalous blue interference color.

Mineral Association

A common rock forming mineral, chlorite is associated with a wide range of minerals ranging from clays, micas, amphiboles, feldspars, and common metamorphic ferromagnesium minerals.

Gluconite

Nature of the Phase

A complex solid solution. $(K, Ca, Na)_{\sim 1.6}(Fe^{3+}, Al, Mg, Fe^{2+})_4 Si_{7.3} Al_{0.7} O_{20}(OH)_4$.

Crystallography

Monoclinic, 2/m, C2/m (1M polytype).

$$Z = 1$$
$$á = 1.59 - 1.61$$
$$â = 1.61 - 1.64$$
$$ã = 1.61 - 1.64$$

Increase Fe^{3+} content of the mineral causes increase in refractive indices.

Plane Polarized Light

Form and Relief: Fine grained scales, often occupying intergranular spaces in sandstones and siltstones; may form small laths in impure limestones. Relief low (Fig. 14.13).

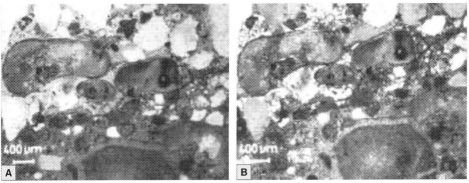

(Courtesy: D.M. Banerjee)

Fig. 14.13 Photomicrographs of glauconite. **A.** Plane polarized light; **B.** Crossed polars.

Color and Pleochroism: Green color, becomes darker as Fe^{3+} substitution increases. Pleochroic: á = pale green or green, â = ã = bluish green. á < â = ã.

Cleavage: Perfect basal cleavages {001}.

Under Crossed Polars

Anisotropism: Anisotropic.

Birefringence: Low. Best seen in sections normal to cleavage.

ä: 0.02 – 0.03.

Polarization Colors: Low first order, rarely exceeding yellow. Anomalous in many cases.

Extinction: 0° – 10° normal to cleavages. Negative elongation (length fast) for biaxial positive chlorites and positive elongation (lengths slow) for biaxial negative chlorites.

Orientation: á approx. normal to {001}, â = y, ã ∧ z = ~0° (Fig. 14.14).

OAP: {010}.

Indicatrix: Biaxial negative.

$2V_a$: 0° – 20°.

Sign of Elongation: Cleavage traces are slow.

Dispersion: Moderate dispersion v > r.

Twinning: Common, multiple {001}.

Zoning: Not seen.

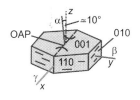

Fig. 14.14 Optic orientation of glauconite.

Alteration

Not common.

Distinguishing Characters

Green color low birefringence and anomalous blue interference colors.

Mineral Association

A characteristic mineral of sandstone of marine origin. Several clay minerals are associated with it.

Apophyllite

Nature of the Phase

Pure phase: $(K,Ca)Ca_4[Si_8O_{20}]$ $(F,OH).8H_2O$.

Crystallography

Tetragonal, 4/m2/m2/m, P4/mnc.

$$Z = 2$$
$$å = 1.535 - 1.543$$
$$ù = 1.534 - 1.542$$

Plane Polarized Light

Form and Relief: Prismatic laths are common. Negative relief (Fig. 14.15).

Color and Pleochroism: Colorless; non-pleochroic.

Cleavage: Perfect basal cleavage {001}, {110} poor.

Under Crossed Polars

Anisotropism: Anisotropic.

Birefringence: Very weak.

ä: $0.000 - 0.003$.

Fig. 14.15 Photomicrograph of apophyllite in plane polarized light.

Polarization Colors: First order gray, commonly anomalous.

Extinction: Parallel.

Indicatrix: Uniaxial positive. Centred figure seen normal to {001} with broad isogyres and practically no isochrones.

Sign of Elongation: Negative elongation.

Dispersion: High, may be anomalous in some cases.

Twinning: Not seen.

Zoning: Not seen.

Alteration

Not common.

Phyllosilicates 295

Distinguishing Characters

Negative relief and anomalous interference colors.

Mineral Association

Zeolites, calcite and prehnite are common associates.

Stilpnomelane

Nature of the Phase

A complex hydrous Mg-Fe silicate. $[K, Na, Ca]_{0-1}$ $[Fe^{3+}Fe^{2+}, Mg, Mn, Al]_2(O, OH, H_2O)_{6-7}[Fe^{3+}, Fe^{2+}, Mg, Mn, Al]_{5-6}Si_8Al(O, OH)_{27}2-4H_2O$.

The solid solution is multicomponent, though the principal replacement is between Fe^{3+} and Fe^{2+}. Higher valent cation's entry is accompanied by increase in O^{2+} at the expanse of $(OH)^-$; often additional H_2O molecules are formed. The prefix ferro- is added to stilpnomelane if Fe^{2+}: Fe^{3+} ration is higher; for reverse cases ferri-prefix is added. Another noteworthy substitution is that of Mg for Fe^{2+} or even Fe^{3+}.

Crystallography

Triclinic, $\bar{1}$, $P\bar{1}$.

$$Z = 8$$
$$á = 1.543 - 1.634$$
$$â = 1.576 - 1.754$$
$$ã = 1.576 - 1.754$$

The refractive indices vary with increase in Fe content in stilpnomelane though influence of Fe^{3+} is more pronounced in comparison to Fe^{2+}.

Plane Polarized Light

Form and Relief: Aggregates of thin plates or scales. Low to moderate relief; relief negative in character for some directions.

Color and Pleochroism: Yellow or brown, at times green. Strongly pleochroic. á = yellow, â = ã = dark brown to black. ã = â > á.

Cleavage: Two sets of cleavages. Perfect along {001}; good along {010}.

Under Crossed Polars

Anisotropism: Anisotropic.

Birefringence: Moderate to high; {010} section shows maximum birefringence, but the basal section shows no birefringence.

ä: 0.03 – 0.110 depending upon the chemical composition. Increases with increasing ferric-ferrous ratio.

Polarization Colors: Brilliant, ranging to middle second order and may be anomalous.

Extinction: Extinction is parallel to cleavage traces.

Orientation: á ∧ z = –7°, â = y, ã ≈ x (Fig. 14.16).

OAP: {010}.

Indicatrix: Biaxial negative; most examples are nearly uniaxial. The basal section is most suited for obtaining a figure, though because of high absorption it may be difficult to view. When seen the figure often resembles a uniaxial figure with several isogyres.

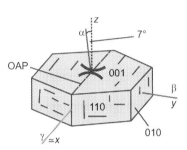

Fig. 14.16 Optic orientation of stilpnomelane.

2V$_a$: Nearly zero.

Sign of Elongation: Positive (length slow).

Dispersion: Strong absorption makes it difficult to asses.

Twinning: Not seen.

Zoning: Faint zoning may be present because of ferric-ferrous ratio.

Alteration

At times alters to a mixture of clays and amorphous iron oxides.

Distinguishing Character

High birefringence and presence of second cleavage are quite characteristics. In general, the mottled appearance or "bird's eye maple" of micas under cross polars is also lacking.

Mineral Association

Stilpnomelane is metamorphic mineral characteristic of low temperature. It is typically found in banded iron formation with other iron oxide minerals. Both greenschists and blueschists may contain stilpnomelane in iron rich protoliths where chlorite, epidote, garnet etc. are common associates.

Prehnite

Nature of the Phase
Nearly a pure phase calc-silicate. $Ca_2[Al, Fe^{3+}](AlSi_3)O_{10}(OH)_2$.

Crystallography
Orthorhombic, 2/m2/m2/m, P2cm, or P2/n.

> $Z = 2$
> $á = 1.612 – 1.637$
> $â = 1.621 – 1.647$
> $ã = 1.639 – 1.673$

Plane Polarized Light

Form and Relief: Sheaf like or fan shaped aggregates are common. Relief moderate. In many instances, radiating sectors or "hourglass" segments habit seen.

Color and Pleochroism: Colorless and non-pleochroic.

Cleavage: Distinct basal cleavage {001}.

Under Crossed Polars

Anisotropism: Anisotropic.

Birefringence: Moderate, best seen in {010} section. Maximum birefringence, is moderate.

$ã$: $0.020 – 0.035$.

Polarization Colors: Anomalous colors. Normal interference colors range up to middle second order.

Extinction: Parallel, often wavy.

Orientation: $á = x$, $â = y$, $ã = z$ (Fig. 14.17).

OAP: {010}

Indicatrix: Biaxial positive. Basal sections yield centred biaxial figure with a few colored rings.

$2V_á$: 64° to 71°.

Sign of Elongation: length-slow (positive elongation).

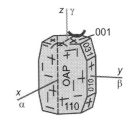

Fig. 14.17 Optic orientation of prehnite.

Dispersion: Usually weak $r > v$; however abnormal crystal sectors may show smaller 2V, possible reversed dispersion ($v > r$ weak to strong) and even crossed bisectrix dispersion.

Twinning: May be present. Resembles microcline twinning.

Zoning: Not common.

Alteration

Not common.

Distinguishing Characters

Calc-silicate mineral association is most characteristic property alongwith a higher birefringence.

Mineral Association

Pumpelleyite is the most common associates along with zeolites in low grade (lower) greenschist facies rocks.

■■■

15
Tektosilicates

- Silica Group
- Feldspar Group
- Nepheline Group
- Zeolite Group

Technical Text Box 15.1

Tektosilicates, or framework silicates, along with phyllosilicates form almost four-fifth of the entire crust of the earth. Many members of these silicates incorporate a large number of lithophile elements in trace amount as well as a number of rare earth elements. These minerals contribute towards the differentiation of igneous rocks; their weathering properties control the quality of soil and of basinal waters of sedimentary rocks.

Each of the four oxygen atoms of every silicon tetrahedra is shared with those in the adjacent tetrhedra to yield a compact framework. The scheme of linkage or, polymerization, gives rise to different groups of tektosilicates. However, it may serve us right to recall a favourite Jim Thompson teaching line: take four molecules of quartz and write the composition as Si_4O_8; now replace one Si by one Al and in order to maintain charge balance add one K to make it $KAlSi_3O_8$! A variety of such substitutions and permutations would yield a large number of tektosilicates.

(Courtesy: S. Sengupta)

A photomicrograph of a euhedral plagioclase phenocryst in basalt exhibiting zoning and albite twinning.

SILICA GROUP

Members of this group forms a large number of minerals of which quartz is the most common. The mineral quartz is ubiquitous in the continental crust as well as on the surface of the earth. A number of varieties has been known since pre-historic times. We shall only describe those varieties that are optically distinguishable.

á-Quartz

Nature of the Phase

Essentially a pure phase. SiO_2. However, most natural quartz contains small (ppm to ppb level) amount of certain elements which give rise to several colored varieties, though color is visible only at hand specimen scale. For example, blue or greysh blue color of quartz from charnockite has been attributed to dissolution of colloidal TiO_2.

Crystallography

Hexagonal, occurs in two enantiomorphic forms; 32, $P3_121$, $P3_221$.

$$Z = 3$$
$$\text{å} = 1.553$$
$$\text{ù} = 1.544$$

Plane Polarized Light

Form and Relief: Anhedral grains of coarse to fine-grained size, also as subrounded to rounded grains; fracture filling veins or geode. In many rhyolite samples quartz phenocrysts are euhedral. Relief poor or even absent (Chapter 3, Fig. 3.2D and E).

Color and Pleochroism: Colorless in standard thin sections. Grains may contain liquid and solid inclusions. Solid inclusions, especially rutile needles impart colors to hand specimens (Tyndall scattering). Non-pleochroic.

Cleavage: Cleavages are normally not seen, though under tectonic pressures, at times, yield prismatic $\{10\bar{1}1\}$ or $\{01\bar{1}1\}$ cleavage fragments. In some deformation experiments prismatic planes also separate out.

Under Crossed Polars

Anisotropism: Yes, but basal sections are isotropic.

Birefringence: Very weak. Best seen in prismatic sections.

ä: 0.009 – 0.011.

Polarization Colors: First order colors, maximum up to light yellow.

Extinction: In most cases, it is not possible to get a good straight edge. The extinction is often non-uniform in quartz because of displacement of optic axis. Such extinction is called undulatory extinction (Fig. 15.1).

Indicatrix: Uniaxial positive. Basal section yields centred figure with broad isogyres set in a grey background. When a very thick section (about 1mm) is viewed the isogyres are narrow and sharp with a large number of isochromes. At the centre of the view a large yellow circle appears. This is caused by the optical activity of quartz; because of this property the crystal is capable of rotating the plane of vibration of waves parallel to the optic axis either clockwise or anticlockwise. A right-hand quartz will show the central color to ascend in order in the Michael Levy chart when the N-S upper polar is rotated clockwise. Reverse would be the case for the left-handed quartz.

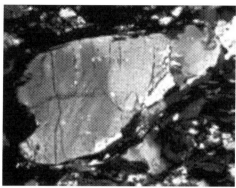

Fig. 15.1 Photomicrograph of a large and fractured quartz porphyroclast from a feldspar gneiss. Note undulose extinction of the grain. Crossed polars.

(Courtesy: A. Joshi)

Sign of Elongation: When elongated, it is length slow (Positive elongation).

Dispersion: Weak; $r > v$ over á.

Twinning: Commonly present but rarely seen under microscope because of similarity in optic orientation of the individuals. Table 15.1 gives details of twin laws.

Table 15.1 Important twin laws of quartz

Name	Twin Axis	Twin Plane
Dauphané Law	z	$\{10\bar{1}0\}$
Brazil Law		$\{11\bar{2}0\}$
Japanese Law		$\{11\bar{2}2\}$
Combined Dauphané-Brazil Law	z	$\{10\bar{1}0\}, \{11\bar{2}0\}$

Zoning: Not present.

Alteration

Not present.

General Text Box 15.1

PROFESSOR CLIFFORD FRONDEL

Institutions like Harvard and MIT are known for housing trend-setting professors who produce trend-setting students. If one were to quote an example from 20th century, for the earth sciences, the names of Martin Buerger and Clifford Frondel would occupy the number one position, and, very worthily followed by Martin Buerger and James B. Thompson Jr. Prof. Clifford Frondel was born in New York City on January 8, 1907, and, stayed there till his high school graduation. For his undergraduate degree, he moved westward to the Rockies for a degree in Geological Engineering from Colorado School of Mines in 1929, but returned to the East Coast to complete Master's from Columbia in 1936 and Ph.D. from MIT in 1939. Prof. Charles Palache took him as his associate at Harvard in 1939 itself, and he remained there till his retirement. He came to occupy the prestigious position of the Curator of Harvard's world famous Mineralogical Museum and Chairman of the Department of Geological Sciences (now known as the Department of Earth and Planetary Sciences). In addition, through assignments in his popular courses, he built up world's largest collection of internal structure of a large number of minerals at Harvard. These structures were built to the scale by means of a large number of coloured plastic balls of different sizes mimicking ions of different elements joined together by thin brass pipes.

His research articles number over 150 with close to 100 being written without any co-author. With Palache and Berman, he produced monumental two volumes of Dana's System of Mineralogy, and single-handedly brought out its Volume 3. He was an early pioneer in the field of mineralogy and geochemistry of uranium and thorium. He is credited with the discovery of 48 new minerals. He was one of the first to synthesize tourmaline, star ruby and sapphire. One of his greatest contributions, is on the piezoelectric properties of quartz. He began these studies when as a civilian he joined Signal Corps of the War Department as well as the Director of research at Reeves Sound Lab during World War II. He was responsible for the construction of quartz oscillator-plates for controlling radio frequencies. These were the precursors to the modern communication system. When NASA began manned mission to the Moon, Prof. Frondel became its advisor. Prior to that he had already established himself as a leading worker on meteorites.

He was frequently honoured for the wealth of his academic output. He received the Roebling Medal of the Mineralogical Society of America. Two minerals have been named after him: (*a*) frondelite, the manganese analogue of rockbridgeite named in 1949 and, (*b*) cliffordite, a uranium tellurite named in 1969.

Contd. ...

Tektosilicates

303

Clifford Frondel was a jovial fellow, never tired of working or enjoying. The story goes that after Apollo 12 mission, when a group of petrologists and mineralogists were studying the samples brought back by the astronauts, the entire team was quarantined as nobody knew how fatal the effects of the lunar samples could be. The quarantine continued for over two weeks during which, the group leader, Frondel, taught them poker, and obviously, robbed them flat! Prof. Clifford Frondel married Judith Weiss, and had a long married life lasting over 53 years. He died on Nov. 12, 2002.

Based on several articles particularly the Buerger presentation of Roebling medal (The American Mineralogist, Vol. 50 March–April 1965) and obituary reference (Geotimes March 2003).

Distinguishing Characters

Undulose extinction is characteristic though lack of cleavage and uniaxial character also confirm the identification.

Mineral Association

Because of its large range of stability, the mineral occurs in a large number of igneous rocks. It is primary mineral of acid igneous rocks such as granites and rhyolites. It is the predominant detrital mineral in sandstones and mudstones, and a major mineral in argillaceous rocks, *e.g.*, shales. Metamorphic derivatives of these igneous and sedimentary rocks such various schists and gneisses contain abundant quartz. In hydrothermal ore deposits quartz is quite common as a gangue mineral.

á-Tridymite

Nature of the Phase

Essentially a pure phase. SiO_2.

Crystallography

Orthorhombic (pseudohexagonal); 222; $C222_1$.

$$Z = 8$$
$$á = 1.469 - 1.479$$
$$â = 1.479 - 1.480$$
$$ã = 1.473 - 1.483$$

Plane Polarized Light

Form and Relief: It occurs in voids and vugs of siliceous volcanic rocks. The crystals are euhedral though normally of small grain size. The most common habit is prism and pinacoid forms terminated by basal pinacoids. Relief strongly negative.

Color and Pleochroism: Colorless. Non-pleochroic.

Cleavage: No cleavages.

Under Crossed Polars

Anisotropism: Anisotropic.

Birefringence: Very weak. Best seen in {100} sections, but may require the help of quarter lambda plate for observataion.

ä: 0.003.

Polarization Colors: First order grey color.

Extinction: Parallel.

Orientation: á = y, â = x, ã = z (Fig. 15.2).

OAP: {100}.

Indicatrix: Biaxial positive. Difficult to observe.

$2V_{\mathrm{a}}$: 40° – 90°.

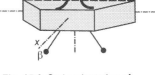

Fig. 15.2 Optic orientation of á-tridymite.

Sign of Elongation: When elongated, it is length slow (Positive elongation).

Dispersion: Weak; $r > v$.

Twinning: Commonly present; contact twins resulting in wedge-shaped grains.

Zoning: Not present.

Alteration

Not present.

Distinguishing Characters

High negative relief and twinning are most characteristics.

Mineral Association

á-tridymite occurs as cavity and fracture fillings in silica rich volcanic rocks where it is associated with quartz and cristobalite. Reported from stony meteorites.

Tektosilicates

α-Cristobalite

Nature of the Phase

Essentially a pure phase. SiO_2.

Crystallography

Tetragonal (pseudoisometric), occurs as enantiomorph; 422; $P4_12_12$ or $P4_32_12$.

$$Z = 4$$
$$å = 1.484$$
$$ù = 1.487$$

Plane Polarized Light

Form and Relief: Often euhedral as octahedra. Relief strongly negative.

Color and Pleochroism: Colorless. Grains may contain liquid and solid inclusions. Non-pleochroic.

Cleavage: No cleavages; sets of curved fractures.

Under Crossed Polars

Anisotropism: Yes, but is nearly isotropic.

Birefringence: Very weak.

ä: 0.003.

Polarization Colors: First order grey color.

Extinction: It is not of much relevance since á-cristobalite is nearly isotropic.

Indicatrix: Interference figure difficult to observe.

Sign of Elongation: Difficult to observe.

Dispersion: Weak; $r > v$.

Twinning: Common on spinel law.

Zoning: Not present.

Alteration

Not present.

Distinguishing Characters

Euhedral grains, negative prominent relief, nearly isotropic nature.

Mineral Association

á-cristobalite is commonly associated with tridymite, quartz, sanidine etc.

Coesite

Nature of the Phase

Essentially a pure phase. SiO_2.

Crystallography

Monoclinic.

$$Z = ?$$
$$á = 1.590 - 1.594$$
$$â =$$
$$ã = 1.597 - 1.604$$

Plane Polarized Light

Form and Relief: It occurs as small anhedral grains usually as minute inclusions within other minerals such as garnet, zircon, omphacite etc. Relief is average positive against quartz (Fig. 15.3).

Color and Pleochroism: Colorless. Non-pleochroic.

Cleavage: No cleavages.

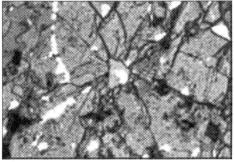

(Courtesy: P. Singh)

Fig. 15.3 Photomicrograph of coesite (?) inclusion (anhedral grain in the centre of view with cross fractures emanating from it) in garnet in plane polarized light.

Under Crossed Polars

Anisotropism: Anisotropic.

Birefringence: Weak.

ä: $0.007 - 0.01$.

Polarization Colors: First order grey color.

Extinction: Not possible to measure.

Indicatrix: Not possible to measure.

Sign of Elongation: Not possible to measure.

Dispersion: Not possible to measure.

Tektosilicates

Twinning: Not present.

Zoning: Not present.

Alteration

Commonly alters to quartz.

Distinguishing Characters

Alteration to quartz producing radial cracks (Fig. 15.3), polygonal quartz grain association.

Mineral Association

á-cristobalite is commonly associated with tridymite, quartz, sanidine etc.

FELDSPAR GROUP

Feldspars are important rock-forming minerals. The members of this group show a wide range of chemical composition and crystal chemical behaviour that has been correlated with environment of rock formation. The plagioclase series of this group has been employed in the classification igneous rocks because of its ubiquitous nature and its capacity to reflect the bulk composition of rocks. Despite the fact that ternary feldspars are not common in nature, it is convenient to describe the composition of feldspar minerals in terms of a ternary system, $NaAlSi_3O_8$ (Ab) – $KAlSi_3O_8$ (Or) – $CaAl_2Si_2O_8$ (An). This system is shown in Fig. 15.4 along with nomenclature of major members of this group. You may notice that within the composition field of Fig. 15.4 a large part is not accessible to natural or synthetic feldspars. Hence, it is possible to have a two fold classification of feldspars.

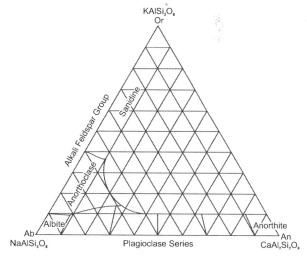

Fig. 15.4 Ternary system, $NaAlSi_3O_8$ (Ab) – $KAlSi_3O_8$ (Or) – $CaAl_2Si_2O_8$ (An).

Feldspar minerals respond to rising temperature by rearranging the distribution of tetrahedral occupancy between one Al and three Si atoms in four-membered rings so characteristic of feldspar structure. There is a complete ordering of Al in $T_{1,0}$ sites at temperature below 300°C; increasing degree of disorder takes place with the rise in temeratuire such that above 550°C there is complete disordering. Information about the degree of disordering comes directly from x-ray and neutron diffraction studies as well as by spectroscopic techniques such as NMR and IR. Inferences from optical observations regarding state of ordering may also be made. On the basis of ordering the feldspars may be classified as:

(*i*) homogenous high-temperature disordered feldspars;
(*ii*) ordered feldspars commonly two phase assemblage on microscopic or submicroscopic scale.

Most natural samples of feldspars contain a small amount of barium with BaO ranging from 0.1 to 2%.

ALKALI FELDSPAR GROUP

Nature of the Phase

Essentially a binary solid solution between potassium and sodium feldspar, *i.e.*, $KAlSi_3O_8$ – $NaAlSi_3O_8$. On the basis of internal structure the group is further divided into three series, *viz.*, the sanidine – high albite series, the orthoclase – low albite series, and the microcline – low albite series.

The Sanidine – High Albite Series

Nature of the Phase

In nature only the potassium end-member exists, *viz.*, $KAlSi_3O_8$; however, it is possible to synthesize the complete solid solution series: $KAlSi_3O_8$ – $NaAlSi_3O_8$. The K-end-member is called high sanidine and the Na-end-member is called high albite (Ab).

Crystallography

High sanidine is monoclinic, but high albite triclinic; monoclinic – triclinic conversion occurs at Ab_{63}.

		High Sanidine	Sanidine	Anorthoclase	High Albite
á	=	1.518 – 1.521	1.518 – 1.524	1.524 – 1.526	1.526 – 1.526
â	=	1.523 – 1.525	1.522 – 1.524	1.529 – 1.532	1.532 – 1.533
ã	=	1.524 – 1.526	1.522 – 1.530	1.530 – 1.534	1.534 – 1.541

Tektosilicates

Plane Polarized Light

Form and Relief: Euhedral square shaped or elongated parallel to x-axis. Relief poor to moderate but negative against araldite (cementing material) or quartz.

Color and Pleochroism: Colorless. Non-pleochroic.

Cleavage: Perfect basal and pinacoidal; {001}is comparatively sharper than {010}. Partings along {100}, (110), {$\bar{1}$10}, and {$\bar{2}$01}.

Under Crossed Polars

Anisotropism: Anisotropic.

Birefringence: Very weak.

ä: 0.006 (High Sanidine); 0.005 – 0.006 (Sanidine); 0.006 – 0.007 (Anorthoclase); 0.007 – 0.008 (High Albite).

Polarization Colors: First order colors, maximum up to first order grey.

Extinction: Extinction angle is low though exact value depends upon orientation. For example {001} cleavage seen in {010} form of sanidine, the extinction angle is about 5°, but {001} cleavage seen on {100} gives parallel extinction.

Orientation: For sanidine the orientation is á $\wedge$ x = 5°, â = z, ã $\wedge$ y = –20°, which gradually changes to á $\wedge$ x = 10°, â $\wedge$ z =20°, ã $\wedge$ y = 5° for anorthoclase (triclinic); a slight increase of the angles occur as the composition moves towards high albite. On the other hand, high sanidine shows á $\wedge$ x = 5°, â = y, ã $\wedge$ z =–20°. For graphic relationship *see* Fig. 15.5. The orientation of albite is given in the description for plagioclase.

Indicatrix: Biaxial negative. The basal section yields a flash figure; the pinacoid sections give diffused figure. High sanidine yields the diffused figure on {001} and a flash figure on {010}.

OAP: The OAP is normal to {010} at the K-end of the composition, but gradually becomes nearly parallel to it towards the Na-end.

$2V_a$: For K-end-members it is 18° gradually becoming 54° at the Na-end.

Sign of Elongation: Sign of elongation is negative; cleavage traces are fast.

Dispersion: Weak; $r > v$ except for sanidine which shows $v > r$.

Twinning: Sanidine commonly exhibits Carlsbad twinning which is simple law with only two individuals twinned along c-axis. Anorthoclase being triclinic shows combination of albite and pericline twin laws. However, unlike microcline, the grid pattern is shown by {100} sections. {001} and (010) sections reveal only one law, the former the albite, and the latter pericline.

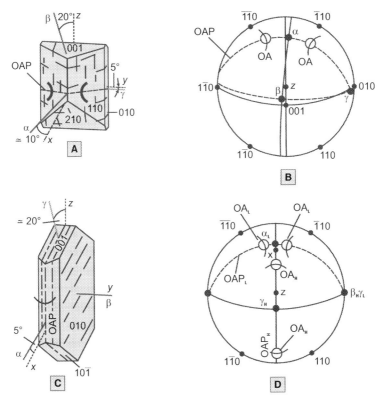

Fig. 15.5 A. Optic orientation of anorthoclase; **B.** Stereogram of the optic orientation of anorthoclase; **C.** Optic orientation of high sanidine; **D.** Stereogram of high and low sanidine.

Zoning: Not present.

Alteration

Common with sericite and clays as the alteration products.

Distinguishing Characters

The occurrence of sanidine as phenocrysts in lavas of silicic composition distinguishes it clearly. Its $2V_a$ and extinction angle are smaller than those of lower T K-feldspars.

Mineral Association

The members of this series are rare and occur only in high K-lavas, or as a product of contact metamorphism in xenoliths within basalts. Quartz is the most common associate.

The Orthoclase – Low Albite Series

Nature of the Phase

This series exists as one of complete miscibility between the K- and Na- end members at high temperature. At lower temperature the solid solution breaks down resulting in nearly pure end-members. The initial intermediate members, upon cooling become unmixed and show lamellar structure called perthite. A number of descriptive names are used in the literature for various geometric and crystallographic orientations of lamellae, *e.g.*, cryptoperthite, microperthite etc. (Fig. 15.6). Pure K-phase and pure Na-phase are called adularia and pericline respectively provided they preserve the high temperature forms. The sodic end-member invariably may also contain Ca, Ba, Sr or even Mg and Fe^{2+}; the tetrahedral Al in the structure of this feldspar may also be replaced by Fe^{3+}.

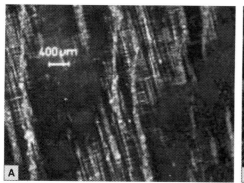
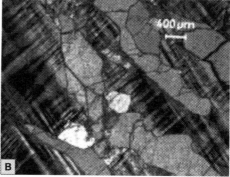

(Courtesy: Department of Geology Collection, DU)   (Courtesy: Department of Geology Collection, DU)

Fig. 15.6 Examples of perthites. **A.** String perthite; **B.** Graphic perthite.

Crystallography

Monoclinic with domains of triclinic in orthoclase, low albite is triclinic. $\bar{1}$; $C\bar{1}$.

	Orthoclase (Or_{100}-Or_{85})	Low Albite (Or_{20}-Or_{0})
á =	1.518 – 1.520	1.527 – 1.530
â =	1.522 – 1.524	1.531 – 1.533
ã =	1.522 – 1.525	1.536 – 1.540

Plane Polarized Light

Form and Relief: Anhedral grains are common (Fig. 15.7). Adularia is euhedral square shaped or elongated parallel to *a*-axis. Relief poor to moderate but negative against araldite (cementing material) or quartz.

Under Crossed Polars

Anisotropism: Anisotropic.

Birefringence: Very weak.

δ: 0.005 [Orthoclase ($Or_{100} - Or_{85}$)] – 0.009 [Low Albite ($Or_{20} - Or_{0}$)].

Polarization Colors: First order colors, maximum up to first order grey (Fig. 15.7).

Extinction: Extinction angle values depend the Na-content of the mineral; highest value, 19°, is for albite end-member and decreases as K fraction increases, finally measuring only 6°. These values are shown by {010} section that prominently shows {001} cleavages.

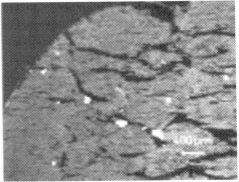

Fig. 15.7 Photomicrograph of orthoclase in crossed polars.

(Courtesy: Department of Geology Collection, DU)

Orientation: For orthoclase: $\alpha \wedge x = 14°$ to 6°, $\beta \wedge z = -13°$ to $-21°$, $\gamma = y$ (Fig. 15.8); for low albite see next section on plagioclase.

Indicatrix: Orthoclase shows a biaxial negative figure whereas low albite shows a biaxial positive figure.

2V: $2V_\alpha$ for orthoclase is 35° – 50°; for low albite $2V_\gamma$ is 90° – 85°.

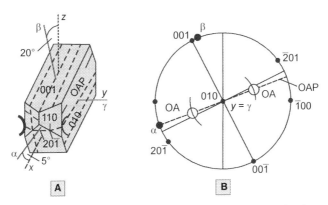

Fig. 15.8 A. Optic orientation of orthoclase; B. Stereogram of optic orientation of orthoclase.

Sign of Elongation: Cleavage traces are fast for orthoclase; albite will be described latter along with plagioclase.

Dispersion: Moderate for orthoclase, $r > v$; weak dispersion in low albite, $v > r$.

Sign of Elongation: Cleavage traces are fast for orthoclase; albite will be described latter along with plagioclase.

Dispersion: Moderate for orthoclase, $r > v$; weak dispersion in low albite, $v > r$.

Twinning: All members of this series commonly exhibit Carlsbad twinning (Fig. 15.9A) which is simple law with only two individuals. The interesting observation is concerned with the orientation of the section with reference to the composition plane. If the section is parallel to b-axis both twins illuminate with different birefringence but simultaneously extinguish when the composition plane is parallel to either crosswire. It is not easy to see this twinning when the section is cut parallel or normal to c-axis. Baveno twins (Fig. 15.9B) are also observed in this series where the composition plane is seen as a

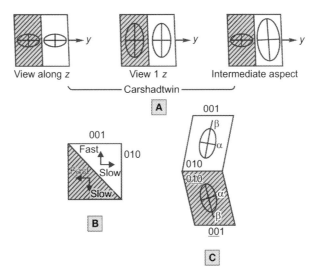

Fig. 15.9 Illustrations of optical relationships in Orthoclase. **A.** Carlsbad; **B.** Baveno; **C.** Manebach.

trace along one of the diagonals of nearly square prism faces. In this case the striking difference between the two individuals of Baveno twins is the reversal of fast and slow directions. Manabach twins (Fig. 15.9C) are recognized on the basis of differences in birefringence.

Zoning: Not seen.

Alteration

Sericite and kaolinite are the most common alteration products.

Distinguishing Characters

Quartz is similar to these feldspars when they are not twinned. Biaxial character and cleavages may be used as distinguishing character from quartz. Plagioclases are normally multiple individuals when twinned.

Mineral Association

In felsic igneous rocks members of this series are associated with quartz, plagioclase, or nepheline.

The Microcline – Low Albite Series

Nature of the Phase

Mineralogically this series is represented by a variety of perthites where host is K-rich end-member and the lamellae are Na-rich. Some name in common practice are microcline perthite, microcline microperthite, cryptoperthite etc. K-rich and Na-rich end-members, *i.e.*, microcline and albite may also form individual crystals as well with both carrying no more than An_{05} of anorthite.

Crystallography

Triclinic, $\bar{1}$, P$\bar{1}$.

$Z = 4$		Microcline (Or$_{100}$ – Or$_{92}$)	Low Albite (Or$_{20}$ – Or$_0$)
á	=	1.514 – 1.516	1.527 – 1.530
â	=	1.518 – 1.519	1.531 – 1.530
ã	=	1.521 – 1.522	1.536 – 1.540

Plane Polarized Light

Form and Relief: Anhedral grains are common. Relief poor to moderate but negative against araldite (cementing material) or quartz.

Color and Pleochroism: Colorless. Non-pleochroic.

Cleavage: Perfect basal and pinacoidal; {001} is comparatively sharper than {010}.

Under Crossed Polars

Anisotropism: Anisotropic.

Birefringence: Very weak.

ä: 0.007 [Microcline (Or$_{100}$ – Or$_{92}$)] – 0.009 [Low Albite (Or$_{20}$ – Or$_0$)].

Polarization Colors: First order, maximum up to first order grey.

Tektosilicates

Extinction: Extinction angle on sections parallel to {010} give about 5° value, however, on {001} sections the angle rises to 15°.

Orientation: For microcline: á ∧ x = 18°, â ∧ z = –18°, ã ∧ y = 18° (Fig. 15.10); for low albite the orientation is described in the next section.

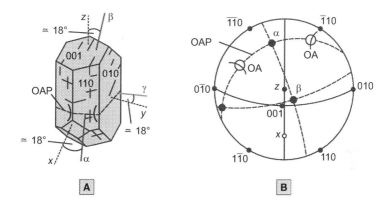

Fig. 15.10 A. Optic orientation of microcline; **B.** Stereogram of optic orientation of microcline.

Indicatrix: Microcline shows a biaxial negative figure whereas low albite shows a biaxial positive figure. Interpenetration twins cause hindrance in obtaining a good figure.

2V: For microcline: $2V_{à} = 66° – 88°$; for low albite $2V_{à} = 90° – 85°$.

Sign of Elongation: Cleavages are length fast.

Dispersion: Moderate in microcline with $r > v$, weak in low albite $v > r$.

Twinning: Microcline is invariably twinned polysynthetically on two laws, *i.e.*, albite and pericline. Because twin planes of both are normal to each other, the repeated twinning give rise to well known quadrille pattern or also called crosshatched twinning (Fig. 15.11A). Albite exhibits polysynthetic twinning on albite and pericline laws but the lamellae are broad and few, therefore quadrille pattern is not present. In addition, twins may also reveal twinning on Carlbad law; which simply is shown by albite laths as half dark grain and the other half exhibiting more than one individuals; upon rotation the dark half becomes bright revealing, just as in the former case, more than one albite individuals.

Zoning: Not seen.

Exsolution: It has already been mentioned that like pyroxenes, alkali feldspars also show phase separation during cooling giving rise to characteristic intergrowth textures

(Courtesy: Geology Department Collection, DU)

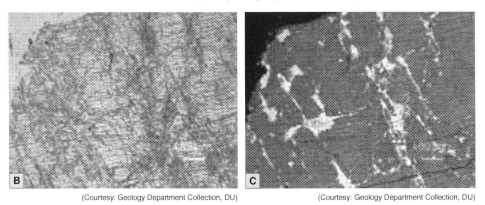
(Courtesy: Geology Department Collection, DU)   (Courtesy: Geology Department Collection, DU)

Fig. 15.11 Photomicrographs of a microcline. **A.** A typical cross-hatched microcline under crossed polars; **B.** With two sets of good cleavages and stringer perthite texture that exsolved Na-rich feldspar along two preferred crystallographic directions under plane polarized light; **C.** Crossed polars.

called perthites. Microline-low albite series exhibits such textures conspicuously. Emmons (Geol. Soc. Amer. Mem. No. 5) shows some excellent illustrations. Fig. 7.7C, in this book, gives such intergrowth example. Figs. 15.11B and C bring out the relationship between crystallographic characters and separation of phases.

Alteration

Sericite and kaolinite are the most common alteration products.

Distinguishing Characters

Quartz is similar to these feldspars when they are not twinned. Biaxial character and cleavages may be used as distinguishing character from quartz. Plagioclase are normally multiple individuals when twinned.

Tektosilicates

Mineral Association

In felsic igneous rocks members of this series are associated with quartz, plagioclase, or nepheline. These members are fairly common in metamorphic rocks both low grade and also in high grade.

Plagioclase Feldspar

Nature of the Phase

By plagioclase feldspar we mean the low temperature polymorph members of albite $(NaAlSi_3O_8)$ – anorthite $(CaAl_2Si_2O_8)$ series, which occur in nature in abundance. The solid solution series is non-ideal and is complicated by NaSi ⇔ CaAl coupled substitution. There are three miscibility gaps in the series: peristerite $(An_2 – An_{16})$, Bøgglid $(An_{48} – An_{58})$, and Huttenlocher $(An_{70} – An_{90})$ as shown in Fig. 15.12. Because of the submicroscopic nature of the grains falling within these gaps, the perthite like intergrowth is not seen normally, rather an apparent continuity of composition is seen. Towards the albite end of the series some amount of K is also observed in the mineral analysis of many albite examples.

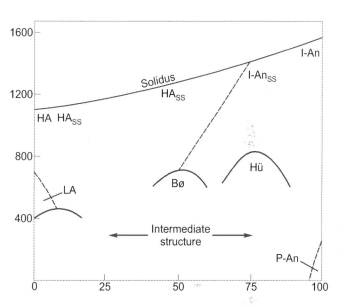

Fig. 15.12 Schematic illustration of variation in structure across the plagioclase series. HA – high albite, I-An – bo anorthite, LA – low albite, P-An – primitive anorthite, Pe – intergrowth, Bø – Bøggild intergrowth, Hu – Hüttenlocher inter (After Smith, J.V. and Brown, W.L., 1988. Feldspars Minerals, 1. Springer-Verlag).

Crystallography

Triclinic, $\bar{1}$, $P\bar{1}$.

		Low Albite (An_0)	Anorthite (An_{100})
Z = 4			
á	=	1.527 – 1.530	1.577
â	=	1.531 – 1.530	1.585
ã	=	1.536 – 1.540	1.590

There is an excellent correlation between the composition of plagioclase and its refractive indices (Fig. 15.13). If you compare these values with those given for the high-temperature polymorphs you would notice the closeness of these values. Hence, irrespective of the original structural state of plagioclase, the refractive index method

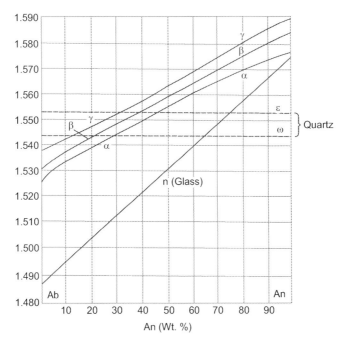

Fig. 15.13 Correlation of refractive indices of plagioclase with its anorthite content.

can be used to estimate composition with reasonable precision. From the chart of Fig. 15.13 it is obvious that á values of plagioclase at the albite-end are less than that of ù of quartz; hence, the plagioclase of low An content (< 30%) would form a negative relief against quartz; then for An values of 30 to 40% the plagioclase relief would merge with that of quartz (no relief); finally, for all those plagioclases that have 70% or more An the relief against quartz would be higher. Hence, a simple Becke line test would yield valuable compositional information. Cleavage fragments of plagio-clase are easily extractable from a rock chip, and, therefore, are commonly used for refractive index determination. It is easy to recognize {001} fragments; when viewed under binocular microscope such fragments appear somewhat rectangular because of intersections of bounding prismatic faces, and also show traces of {010} with albite twin lamellae. On the other hand, {010} are a bit angular because of {100} intersections and no albite twins would be noticed. But irrespective of which cleavage fragment you pick, determination of á will be hardly affected. Modern spindle stages yield an accuracy of 0.001 in á so that composition predictions may have not more than ± 2%.

The refractive index determination does not help if the plagioclase is zoned (Fig. 15.14) and has exsolved lamellae; even EPMA may not be of much help if the zones and lamellae are thin in comparison to beam width. In such cases, it is best to extract a grain and melt it to a glass[1] (DHZ, pp. 445); now the refractive index is easy to determine, particularly, if a yellow light is used. The chart is included in the figure given (Fig. 15.13).

(Courtesy: S. Sengupta)

Fig. 15.14 Photomicrograph of a zoned plagioclase under crossed polars.

Plane Polarized Light

Form and Relief: Anhedral grains are common. Phenocrysts in lavas of all composition may be euhedral. Porphyroblasts in gneisses may be augen shaped. Relief negative to moderate.

Color and Pleochroism: Colorless. Non-pleochroic.

Cleavage: Perfect basal and pinacoidal; {001} is comparatively sharper than {010}.

Under Crossed Polars

Anisotropism: Anisotropic

Birefringence: Very weak.

ä: 0.010 [Low Albite (An_0)] – 0.013 [Anorthite (An_{100})].
Although birefringence varies as An content varies but no systematic relationship has been observed.

Polarization Colors: First order colors, maximum for low albite is up to first order grey, for anorthite a deep yellow color may be seen in cross polars.

Extinction: Extinction angle in plagioclase reveals its chemical composition very nicely. Even without the use of universal stage it can be used to get a reasonable estimate of the anorthite content. The method known after Michel-Lévy uses the fact that the plagioclase laths show twin planes invariably parallel to their length, and commonly, it is an albite twin, *i.e.*, the twin plane is {010}. Extinction angle measured from this plane systematically varies as the composition of the grain changes from An_0 to An_{100}. A grain may not be cut exactly normal to {010} so will yield lower values of extinction angle. If the grain is cut exactly normal to {010} then the trace of the twin plane (coincident with composition plane) will be hairline sharp. This should coincide with its fast vibration

---

1. Deer, Howie and Zussman, 1992, *An Introduction to the Rock-Forming Minerals*, ELBS edition.

direction except in extreme Ca rich plagioclases where this direction is the slow direction. The measurement for the extinction angle for the purpose of determining the An content begins by aligning the composition plane along N-S crosswire, noting the stage reading, then clockwise rotating it to get the nearest extinction position of one of the individuals. The new reading is noted which yields the extinction angle; after the stage is returned to the initial position, a similar extinction angel is estimated by anti-clockwise rotation. If the grain selection was correct both values of the extinction angles should be same or differ by no more than 2°. The average is plotted against the graph of Fig. 15.15 to obtain the An content of the grain. In order to make sure that you are using the maximum value of this angle, repeat this exercise for different suitably selected grains, and, then use the maximum value.

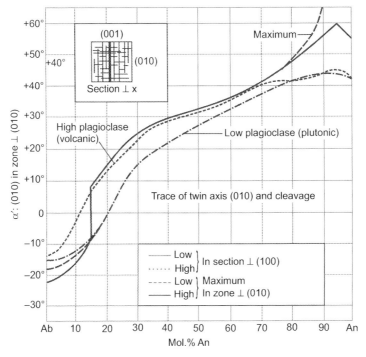

Fig. 15.15 Extinction angles, á' ∧ {010} in the zone ⊥ [010] for high- and low-temperature plagioclase. (After Deer, Howie and Zussman, 1992, *An Introduction to the Rock-Forming Minerals*, ELBS edition).

The value obtained from the above method can be crosschecked by another method after Heinrich. In this method Carlsbad – albite combination twin is used. Of course, like the previous method here, also, the given grain should be a plagioclase lath with its fast direction parallel to Carlsbad composition plane. The values obtained from

Tektosilicates

closckwise and anticlockwise rotations of an albite twin individual within one Carlbad individual should be averaged. Repeat this procedure for the second Carlsbad individual. The plot of Fig. 15.16 uses smaller of the two Carlsbad values as ordinate and the larger

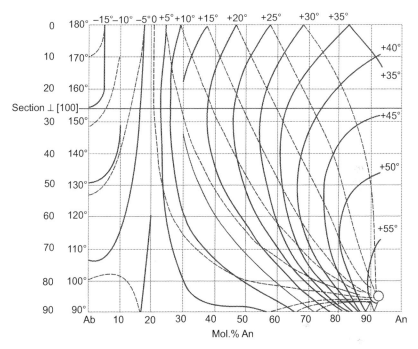

Fig. 15.16 Combined Carlsbad-albite twin method for An determination in plagioclases (After Tobi, A.C. & Kroll, H., 1975, *Amer. J. Sci.*, 275, 731-6).

value is contoured within the plot. You will notice that Michel-Lévy method would yield two values for An content of less than 20% and the Heinrich method yields four values. A general idea about the refractive index obtained from the Becke Line will help you to get to the right combination. As noted above the pure albite end-member has negative relief against quartz.

Orientation: The plagioclase series is triclinic, hence, none of the vibration directions is coincident with any crystal axis. The angular relationship between the two sets of directions changes from An_0 to An_{100}, which, fortunately, allows us to predict the composition of the grain through the measurement of extinction angle. Fig. 15.17 gives the graphic description of this variation.

The optic orientation of a given plagioclase is determined by means of universal stage. However, the current practice is to use spindle stage, which according to several workers yields better results than the U-stage method. The optic orientation is correlated

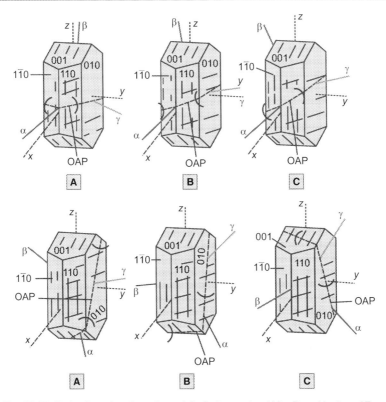

Fig. 15.17 Optic orientation of members of plagioclase series. (After Deer, Howie and Zussman, 1992, *An Introduction to the Rock-Forming Minerals*, ELBS edition).

with the crystallographic data obtained by precision method, finally, the two sets of data are correlated.

Indicatrix: Biaxial, but the positive and negative character is dependent upon the composition of the grain. Because of the complications in the character of solid solution the sign changes more than once. OAP orientation also changes as the composition changes. Fig. 15.18 gives the relationship.

2V: 2V is large and is composition dependent, as well as structural state dependent; as mentioned above the variation is not straight-line relationship. In low T series, abite has $2V_a$ of about 75° (+), whereas in high T series $2V_a$ is 50° (–). Moreover, because of the twinning, the optic figure is not easily obtained.

Dispersion: Weak $r > v$ as well as $v > r$.

Twinning: Present. Albite, pericline, and Carlsbad laws are common. Albite and pericline twins are polysynthetic (Fig. 15.19). The nature of twinned grains is, to some extent, controlled by its mode of occurrence. In plutonic rocks, the composition plane {010} is

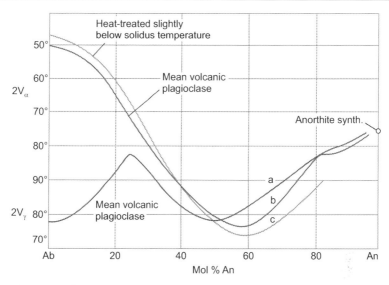

Fig. 15.18 Correlation of $2V_a$ of plagioclase with its anorthite content.

(Courtesy: Geology Department Collection, DU)   (Courtesy: Geology Department Collection, DU)

Fig. 15.19 Photomicrographs of a plagioclase grain. **A.** Exhibiting albite twinning under crossed polars; **B.** Showing both pericline and albite twinning under crossed polars.

the most flat plane (broadest), but the grains are elongated along [001] direction. In volcanic rocks, the grains are elongated along [100] direction. Metamorphism tends to erase twinning so that are fewer twin lamellae mostly on albite law.

Zoning: No zoning in authigenic sedimentary and in metamorphic plagioclases. Igneous plagioclases are zoned (Figs. 15.14 and title page). The normal zoning is Ca-rich core to Na-rich rim. But in many cases oscillatory zoning (composition alternating or irregularly changing between Ca- and Na-rich) has been noticed.

Alteration

Common. Ca-rich variety tends to alter easily as compared to Na-rich one. Sericite and clays are common product and often appear as forming light color dust over the grain. This should not be confused with minute iron oxide particles that cover many igneous plagioclase grains; these are dark colored. Fine-grained aggregate of epidote (or clinozoisite) also clouds the plagioclase grains at times. This process has been described as *saussuritization.*

Distinguishing Characters

Polysynthetic twinning, low relief and low birefringence are the most distinguishing properties that make plagioclase conspicuous.

Mineral Association

All common igneous rocks contain plagioclase. In fact, plagioclase composition is used in rock classification and nomenclature. In compatible minerals like quartz and Mg-rich olivine and pyroxenes occur with plagioclase in different rocks. In sedimentary rocks, it can be produced as a part of sedimentation and diagenesis as well as detrital grains. Common in most regional and contact metamorphic rocks. Alkali feldspars, muscovite and biotite are also common associates.

NEPHELINE GROUP

Nature of the Phase

Most natural occurrences of nepheline have composition of approximately close to $Na_3KAl_4Si_4O_{16}$; this composition is favoured by structural peculiarities of the alkali ion site in nepheline; a quarter of this site are larger than the rest thus allowing K to comfortably replace Na. At high temperature a complete series has been synthesized between $NaAlSiO_4$ and $KAlSiO_4$, as well as partial solid solution between this composition and albite-anorthite.

Crystallography

Hexagonal, 6, $P6_3$.

$$Z = 2$$
$$ù = 1.529 - 1.549$$
$$å = 1.526 - 1.544$$

Since the solid solution is multi-site, the increase in refractive indices cannot be uniquely correlated with composition.

Plane Polarized Light

Form and Relief: It may occur as euhedral grains though subhedral and anhedral habits are common particularly in the plutonic rocks. Moderate negative relief (Figs. 15.20A and B).

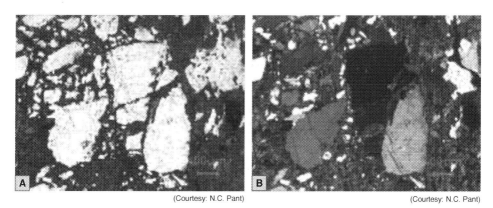

(Courtesy: N.C. Pant) (Courtesy: N.C. Pant)

Fig. 15.20 Photomicrographs of nephaline grains. **A.** Plane polarized light; **B.** Crossed polars.

Color and Pleochroism: Colorless, often appear glassy, in many examples, grains are clouded by inclusions and/or alteration.

Cleavage: Weak cleavages $\{10\bar{1}0\}$ and $\{0001\}$ commonly not seen under microscope.

Under Crossed Polars

Anisotropism: Anisotropic, but the basal sections are isotropic.

Birefringence: Low, seen best in prismatic sections.

ä: $0.003 - 0.005$.

Polarization Colors: Basal sections remain dark; prismatic sections show first order white.

Extinction: Parallel.

Indicatrix: Uniaxial negative. Centred figure is seen on basal sections with broad diffused isogyres.

Sign of Elongation: Negative, length fast.

Dispersion: Weak.

Twinning: May be present with twin planes as $\{100\}$, $\{335\}$ and $\{112\}$.

Zoning: Albite (silica) rich core and Na-rich rims have been observed.

326   Optical Mineralogy

Alteration

Easily altered to other feldspathoids, clays, zeolites etc.

Distinguishing Characters

Untwinned uniaxial grains are quite characteristic.

Mineral Association

Because of its structural and chemical peculiarities, nepheline occurs with a large number of minerals some of which are quite rare. However, one does not expect it to occur with quartz. It is a characteristic mineral of alkali igneous rocks, and, is common in both plutonic and volcanic rocks. Alkali pyroxenes and alkali amphiboles are common minerals associated with nepheline in such rocks. Albite and other feldspathoids are common associates. In igneous rocks formed by hybridization of carbonate country rocks, calcic pyroxenes and calcic amphiboles are common. It is intergrown with feldspars in such metamorphic rocks that are formed by alkali metasomatism.

Leucite

Nature of the Phase

Nearly a pure phase, $KAlSi_2O_6$.

Crystallography

Tetragonal, 4/m, $I4_1/a$; above 625°C at room P, it gets converted to isometric polymorph.

$$Z = 16$$
$$\text{ù} = 1.508 - 1.511$$
$$\text{å} = 1.509 - 1.511$$

Both ù and å are close in values hence, leucite is also called pseudo-isometric.

Plane Polarized Light

Form and Relief: When leucite occurs as phenocrysts in lava, it occurs as euhedral pseudoisometric crystals with {112} form that resembles trapezohedral forms of isometric system. Under microscope, the grains appear as octagons. Relief is high but negative.

Color and Pleochroism: Colorless and non-pleochroic.

Cleavage: Not seen.

Tektosilicates

Under Crossed Polars

Anisotropism: Present but basal sections are isotropic.

Birefringence: Very low.

ä: 0.001.

Polarization Colors: First order grey.

Extinction: Parallel often wavy.

Indicatrix: Uniaxial positive, but the figure is not easy to obtain because the birefringence is very low.

Dispersion: Moderate.

Twinning: Present but seen with great difficulty because of very low birefringence. Three sets of polysynthetic twins intersecting at $60° – 120°$ angle.

Alteration

Alters to pseudoleucite around rim. Pseudoleucite is a mixture of nepheline and K-feldspar.

Distinguishing Characters

Characteristic habit, low birefringence, and polysynthetic twinning.

Mineral Association

Potash rich but Si-poor and Na-poor volcanic rocks contain leucite. Sanidine, olivine, nepheline etc. are common associates.

Sodalite

Nature of the Phase

Almost a pure phase. $Na_8Al_6Si_6O_{24}Cl_2$

Crystallography

Isometric. 3m, P3n.

$$Z = 1$$
$$n = 1.483 – 1.487$$

328 — Optical Mineralogy

Plane Polarized Light

Form and Relief: Aggregates of granular and anhedral grains. At times sodalite forms euhedral dodecahedral crystals forming hexagonal outline in thin sections. Relief high negative.

Color and Pleochroism: Colorless and non-pleochric.

Cleavage: Not present.

Under Crossed Polars

Anisotropism: Isotropic. Other properties not applicable.

Zoning: Not seen.

Alteration

Alters to zeolites and clays.

Distinguishing Character

High negative relief and isotropic character.

Mineral Association

Nepheline, alkali feldspars, and other feldspathoid are common associates.

Scapolite

Nature of the Phase

A binary solid solution series between marialite $(Na_4(Al_3Si_9O_{24})Cl$ to meionite $(Ca_4(Al_3Si_9O_{24})CO_3)$.

Crystallography

Tetragonal, $4/m2/m2/m$, $P4_2/m$, $I4/m$.

$$\grave{u} = 1.540 - 1.600$$
$$\mathring{a} = 1.535 - 1.565$$

According to Deer, Howie and Zussman the marialite: meionite content can be estimated fairly well by the formula $(\mathring{a}+\grave{u})/2 = 1.5346 = 0.000507* (Me\%)$.

Plane Polarized Light

Form and Relief: Aggregates of acicular crystals; slender euhedral prismatic laths with low relief.

Color and Pleochroism: Colorless and non-pleochroic; some varieties show pleochroism in yellow or violet; when so å = yellow or violet, ù = colorless; å > ù.

Cleavage: Faint {100} and {110}.

Under Crossed Polars

Anisotropism: Anisotropic, but basal sections are isotropic.

Birefringence: Low to moderate; governed by the amount of Ca substituted in the mineral.

ä: 0.004 – 0.037.

Polarization Colors: First order grey at the sodic-end to middle second order for the Ca-end (Fig. 15.21).

Extinction: Parallel to {100} cleavage.

Indicatrix: Uniaxial negative; sharpness of the isogyres and the number of colored rings will increase with increase in the Ca content of scapolite.

Sign of Elongation: Negative, prismatic grains are lengthfast.

Dispersion: Moderate.

Twinning: Not seen

Zoning: Na-rich core with Ca-rich rim may be observed.

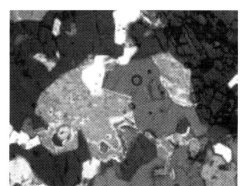

(Courtesy: N.C. Pant)

Fig. 15.21 Photomicrograph of scapolite under crossed polars.

Alteration

May alter to clays.

Distinguishing Characters

Uniaxial negative character with two sets of cleavages.

Mineral Association

An important calc-silicate mineral, scapolite occurs in metamorphic and metasomatic environment. Common minerals occurring with it are sphene, epidote, calcite, Ca-amphiboles. Diopside, grossularite, idocrase etc.

ZEOLITE GROUP

It is perhaps the largest group amongst tektosilicates in terms of number of minerals that are members of this group (>45). For a great part of the history of civilization, these zeolites were known for their decorative values. Since the middle of last century, these minerals and their synthetic equivalents have been employed as cataion exchangers, adsorbents, molecular sieves etc.

Natrolite

Nature of the Phase

Often a pure phase in nature; $Na_2(Al_2Si_3)O_{10}.2H_2O$.

Crystallography

Orthorhombic, 2/m2/m2/m, Fdd2.

$$\alpha = 1.473 - 1.483$$
$$\beta = 1.476 - 1.486$$
$$\gamma = 1.485 - 1.496$$

Plane Polarized Light

Form and Relief: Radial aggregates of fibres or needles, filling amygdales.

Color and Pleochroism: Colorless and non-pleochroic.

Cleavage: Perfect; fibres are formed by {110} cleavage and {010} partings.

Under Crossed Polars

Anisotropism: Anisotropic.

Birefringence: Moderate; best seen on {010}.

δ: 0.012 – 0.013.

Polarization Colors: First order yellow or slightly higher.

Extinction: Parallel.

Orientation: α = x, β = y, γ = z (Fig. 15.22).

Indicatrix: Biaxial positive. Basal sections show centred biaxial figure, difficult to obtain owing to strong fibrous habit.

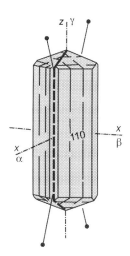

Fig. 15.22 Optic Orientation of natrolite.

Tektosilicates 331

$2V_\gamma$: $58° - 64°$.

OAP: Parallel to {010}.

Sign of Elongation: Fibres are length slow (positive elongation).

Dispersion: Weak, $v > r$.

Twinning: Not seen.

Zoning: Not seen.

Alteration

Normally not altered.

Distinguishing Characters

Fibrous habit, positive elongation and mode of occurrence.

Mineral Association

Other zeolite species.

Chabazite

Nature of the Phase

The composition is highly variable, the general formula is $(Ca, Na, K)(Al_2Si_4)O_{12}.6H_2O$.

Crystallography

Hexagonal.

$$\omega = 1.472 - 1.494$$
$$\varepsilon = 1.470 - 1.485$$

Plane Polarized Light

Form and Relief: Rhombohedral form is common, occurs as cavity filling. Moderate negative relief.

Color and Pleochroism: Colorless and non-pleochroic.

Cleavage: Good rhombohedral cleavage {101}.

Under Crossed Polars

Anisotropism: Anisotropic but basal sections are isotropic.

332 Optical Mineralogy

Birefringence: Low.

δ: 0.002 – 0.010.

Polarization Colors: First order grey.

Extinction: Symmetrical with reference to the cleavage.

Indicatrix: Uniaxial negative.

Sign of Elongation: Positive, may show length character.

Twinning: Common.

Zoning: Present but nature zoning can not be specified.

Alteration

Normally not altered.

Distinguishing Characters

Habit and uniaxial negative character.

Mineral Association

Other zeolites.

Heulandite

Nature of the Phase

Nearly pure phase in nature. $CaNa_2(Al_2Si_7)O_{18} \cdot 6H_2O$.

Crystallography

Monoclinic, m, Cm.

$$\alpha = 1.487 - 1.505$$
$$\beta = 1.487 - 1.505$$
$$\gamma = 1.488 - 1.512$$

Plane Polarized Light

Form and Relief: Tabular laths with low relief.

Color and Pleochroism: Colorless and non-pleochroic.

Cleavage: Perfect {010}.

Under Crossed Polars

Anisotropism: Anisotropic.

Birefringence: Very low best seen on basal pinacoid.

ä: 0.001 – 0.007.

Polarization Colors: First order grey.

Extinction: Cleavage traces show parallel extinction but other sections may show a maximum angle of 33°.

Orientation: á ∧ x = +9° – +33° depending of composition. â = z, ã ∧ y = +8° – +32° (Fig. 15.23).

Indicatrix: Biaxial positive. Interference figure best seen on {010} cleavage fragments.

$2V_a$: 30°.

OAP: {001}.

Sign of Elongation: Cleavage planes {010} are length slow but sections normal to cleavage traces are length fast (negative elongation).

Dispersion: Distinct, $r > v$.

Twinning: Not seen.

Zoning: Not seen.

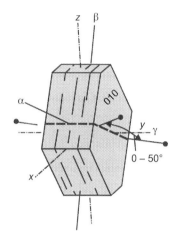

Fig. 15.23 Optic orientation of heulandite.

Alteration

Normally not altered.

Distinguishing Characters

Perfect cleavage, low 2V, and positive character.

Mineral Association

Chabazite and stilbite are common associates; also occurs in bentonite clays.

Stilbite

Nature of the Phase

Multicomponent solid solution. The general formula is $(Ca, Na_2, K_2)(Al_2Si_7)O_{18} \cdot 7H_2O$.

Crystallography

Monoclinic, 2/m, C2/m.

$$á = 1.482 - 1.500$$
$$â = 1.491 - 1.507$$
$$ã = 1.493 - 1.513$$

Plane Polarized Light

Form and Relief: Platy and lath-shaped subhedral grains with low negative relief.

Color and Pleochroism: Colorless and non-pleochroic.

Cleavage: Perfect along {010}; poor along {100}.

Under Crossed Polars

Anisotropism: Anisotropic.

Birefringence: Low, best seen on {010} face.

ä: 0.006 – 0.013.

Polarization Colors: Usually grey of the first order.

Extinction: Parallel to cleavage traces {010}; maximum extinction angle is seen on {010} face up to 32° – 39°.

Orientation: á ∧ z = +33° to +39° depending of composition. â = y, ã ∧ x = 0° to –7° (Fig. 15.24).

Indicatrix: Biaxial negative. The centred optic figure is seen when the section is cut at a high angle to the cleavage. Isogyres are broad and diffused.

2V$_a$: 30° – 49°.

Sign of Elongation: Cleavage traces can be fast or slow depending upon the orientation of the section since â direction lies within the cleavage.

Dispersion: $v > r$.

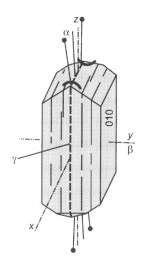

Fig. 15.24 Optic orientation of stilbite.

Tektosilicates

Twinning: Twinned on {001}.

Zoning: Not seen.

Alteration

Not seen.

Distinguishing Characters

Twinning and biaxial negative character.

Mineral Association

With other zeolites like heulandite etc.

Laumontite

Nature of the Phase

Nearly pure phase. $Ca(Al_2Si_4)O_{12}.4H_2O$.

Crystallography

Monoclinic; m; Am.

$$\acute{a} = 1.502 - 1.514$$
$$\hat{a} = 1.512 - 1.522$$
$$\tilde{a} = 1.514 - 1.526$$

Plane Polarized Light

Form and Relief: Prismatic laths and aggregates of radiating fibres. Cavity-filling.

Color and Pleochroism: Colorless and non-pleochroic.

Cleavage: Two sets perfect: {010} and {110}.

Under Crossed Polars

Anisotropism: Anisotropic.

Birefringence: Low, best on {010} cleavage planes.

ä: 0.008 – 0.016.

Polarization Colors: First order grey, becoming first order red in some examples.

Extinction: Inclined on cleavages with variable angles.

Orientation: á ∧ x = +33° – +62° depending of composition. â = y, ã ∧ z = +8° – +40° (Fig. 15.25).

Indicatrix: Biaxial negative; optic figure best seen {100} with broad diffused isogyres.

2V$_a$: 25° – 47°.

OAP: {010}.

Sign of Elongation: Positive. Cleavage traces are length slow.

Dispersion: Strong v > r.

Twinning: Simple twinning on {100} is common.

Zoning: Not seen.

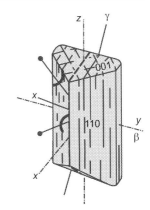

Fig. 15.25 Optic orientation of laumontite.

Alteration
Not seen.

Distinguishing Characters
Fibrous habit with inclined extinction and simple twinning.

Mineral Association
Along with other zeolites.

16

Non-silicates

- Corundum
- Rutile
- Spinel
- Perovskite
- Barite
- Gypsum
- Anhydrite
- Calcite
- Dolomite
- Aragonite
- Apatite
- Monazite
- Xenotime
- Fluorite
- Halite

Technical Text Box 16.1

These minerals are present in minor quantities in rocks, hence, are called accessory minerals. But these are by no means unimportant from the petrology point of view. Their presence, and by the same analogy, their absence may give important clues about the environment of rock formation, including nature of metasomatic fluids. On the other hand, many of them being authigenic to the sedimentary environment are important rock forming minerals, as well.

In addition, some of these may also incorporate radio-elements, thereby, allowing us to date an event in the evolutionary history of rocks.

Non-silicate rock minerals present diverse linkages of coordination polyhedra. For example, in spinal all bonds are isodesmic, *i.e.*, bonds of equal strength. But in carbonates these are anisodesmic since $C^{4+} - O^{2-}$ bond is stronger (CN 3 as triangular) than any other bond in the mineral. Similarly, in sulfates S^{6+} and O^{2-} (planar four fold) are more strongly bounded than other bonds in the mineral. Metal ions, such as Ca, Ba, etc. are weakly bonded to oxygen, hence, are easily replaced by other divalent metal ions.

(Courtesy: P. Singh)

A photomicrograph of marble showing numerous calcite grains; some grains are polysynthetically twinned.

© The Author(s) 2023
P. K. Verma, *Optical Mineralogy*,
https://doi.org/10.1007/978-3-031-40765-9_16

Corundum

Nature of the Phase

Pure phase (Al_2O_3). Trace amounts of Cr, Ti, Fe^{2+}, Fe^{3+} etc. may be present. Different gem varieties originate because of such trace amounts of elements.

Crystallography

Hexagonal, $\bar{3}2/m$, $R\bar{3}c$.

$$Z_{hex} = 6, Z_{rh} = 2$$
$$ù = 1.767 - 1.794$$
$$å = 1.759 - 1.785$$

Substitution of trace elements may alter values of the indices.

Plane Polarized Light

Form and Relief: Large euhedral crystals, showing hexagonal and rectangular sections. Common forms present are hexagonal prism (Fig. 16.1) or steep hexagonal dipyramids yielding barrel form. Prominent relief. Inclusions common.

Color and Pleochroism: Colorless and non-pleochroic; exceptions are Fe-rich varieties, ù = blue and purple, å = blue and green. ù > å.

Cleavage: None, however, basal {0001} parting is common; in several cases rhombohedral partings related to twinning are pre prominently present.

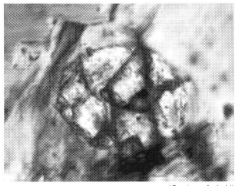
(Courtesy: A. Joshi)

Fig. 16.1 Photomicrograph of corundum in crossed polars.

Under Crossed Polars

Anisotropism: Anisotropic but basal sections are isotropic.

Birefringence: Weak. Best seen in prismatic sections.

ä: 0.008 – 0.009.

Polarization Colors: First order gray and white to pale yellow. However, very high hardness of corundum (9 on Moh's scale), is a handicap in producing exact 30μm thick sections; hence, a comparatively thicker grain will exhibit higher order colors.

Non-silicates

Extinction: Parallel to basal parting and symmetrical to rhombohedral parting.

Indicatrix: Uniaxial negative. Basal sections reveal centred optic figure with broad isogyres set in nearly white background. Because of twinning at times grains show anomalous biaxial character with a 2V up to 30°.

Sign of Elongation: Negative elongation (length fast).

Dispersion: Moderate.

Twinning: Very common; both multiple and simple twinning on rhombohedral planes {101} are very common. This produces lamellar structure (Fig. 16.1), though exolved boehmite may also give rise to such a structure.

Zoning: Common.

Alteration

Alters readily to fine grained, Al-rich minerals, especially margarite or muscovite.

Distinguishing Characters

Extreme relief and weak birefringence with possible parting and lamellar rhombohedral twinning are the most common diagnostic features.

Mineral Association

Intermediate igneous rocks like syenite, nepheline syenite etc. contain corundum; it is also found in many gneisses, especially sillimanite bearing. Muscovite, plagioclase, orthoclase, aluminosilicate polymorphs are common associates.

Rutile

Nature of the Phase

Considered a pure phase; TiO_2. Small amounts of Fe, Ta, Nb, V, Cr, etc. may occur.

Crystallography

Tetragonal, 4/m2/m2/m, $P4_2/mnm$.

$$Z = 4$$
$$ù = 2.605 - 2.613$$
$$å = 2.899 - 2.901$$

Plane Polarized Light

Form and Relief: Two habits are most common, one is euhedral grains characterized by prism traces with pyramid termination at both ends (Figs. 16.2A and B), and the

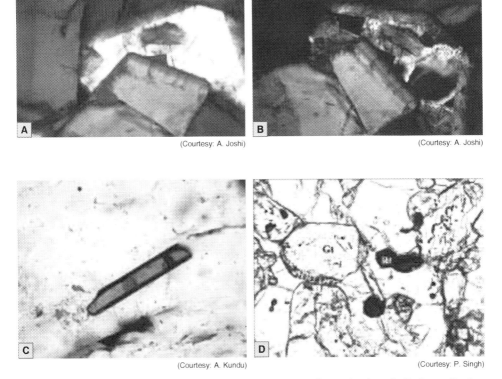

Fig. 16.2 Photomicrographs of rutile. **A.** Plane polarized light; **B.** Crossed polars; **C.** Rutile needles in cordierite. Plane polarized light; **D.** Euhedral rutile grains. Plane polarized light.

second habit is acicular needles (Fig. 16.2C). The latter habit is common when rutile occurs as inclusions in quartz, phlogopite or corundum.

Color and Pleochroism: Deep red brown to yellow brown. Weakly pleochroic. å > ù.

Cleavage: Prismatic cleavage, distinct {110} and fair {100}; however because of high absorption and small grain size it is not easy to observe cleavages. Partings {092} and {011} are also common.

Under Crossed Polars

Anisotropism: Present but basal sections are isotropic.

Birefringence: Very high.

Non-silicates   341

ä: 0.285 – 0.296.

Polarization Colors: High. Usually masked by high absorption (Fig. 16.2D).

Extinction: Parallel.

Sign of Elongation: Positive elongation (length slow). May be difficult to detect.

Dispersion: Very strong (higher than diamond).

Twinning: Very common on {011}. Twinning is often repeated type but simple knee shaped twins are most common (Fig. 16.2D).

Zoning: Not seen.

Alteration

Not seen. Intergrowth with ilmenite in several cases has been observed.

Distinguishing Characters

Red brown color, extreme relief, and extreme birefringence are diagnostic.

Mineral Association

It is a common accessory mineral in a wide range of igneous rocks such as granite as well as anorthosite. It is a common detrital mineral in sedimentary rocks; hence, it is a common mineral in large number of metamorphic rocks. It occurs in rocks characteristic of high temperature like granulites as well as high pressure, *e.g.*, eclogites. Quartz, feldspar, Calcic amphiboles are its common associates.

Spinel

Nature of the Phase

It is a multi-component solid solution. Because of its ferromagnetic properties many new materials with spinel structures have been synthesized. Most members of this group are opaque; although these opaques are very common as rock-forming minerals, but, keeping within the scope of the present book, have not been described here. The common spinel, $MgAl_2O_4$ forms complete solid solutions with hercynite, $FeAl_2O_4$; gahnite $ZnAl_2O_4$; magnesiochromite, $MgCr_2O_4$.

Crystallography

Isometric, 4/m 2/m, Fd3m.

Z = 8	
	n
Spinel	1.716
Hercynite	1.835
Gahnite	1.805
Galaxite	1.920

Plane Polarized Light

Form and Relief: Granular form most common; under microscope euhedral crystals appear as square to triangular sections. Relief high.

Color and Pleochroism: From colorless to various colors depending upon the amount of substitution for Mg. Commonly non-pleochroic though in several cases anomalous pleochroism has been observed.

Spinel	colorless, green, blue, red
Hercynite	dark green
Gahnite	blue-green, yellow-brown
Galaxite	red-brown, black

Cleavage: No cleavage, parting is sometime prominent on {111}.

Under Crossed Polars

Anisotropism: Isotropic.

Twinning: Common on {111} (spinel law) usually simple, sometime multiple.

Zoning: Not seen.

Alteration

Not seen.

Distinguishing Characters

Habit, high relief, isotropism.

Mineral Association

A common mineral in the upper mantle now found in xenoliths occurring within ultrabasic rocks. It is a high grade metamorphic mineral occurring with sillimanite,

Non-silicates   **343**

garnet, corundum, cordierite etc. It is a characteristic mineral of high alumina – low silica metamorphic rocks.

Perovskite

Nature of the Phase

A very complex mineral in terms of its chemical composition. Stoichiometrically, it is $CaTiO_3$, but several substitutions of common and rare elements including radio-elements occur, at times resulting in oxygen vacancies.

Crystallography

Orthorhombic (Pseudocubic), $2/m2/m2/m$, Pcmn.

$$Z = 4$$
$$n = 2.26 - 2.40$$

Plane Polarized Light

Form and Relief: Small euhedral crystals of cubic and octahedral shape. Very high relief.

Color and Pleochroism: Colorless or yellow; practically non-pleochroic.

Cleavage: Not seen

Under Crossed Polars

Anisotropism: Practically isotropic.

Birefringence: Practically nil

Twinning: Twinned on {111} polysynthetically, but rarely observed under microscope.

Zoning: Not seen.

Alteration

Not seen.

Distinguishing Characters

Small euhedral crystals with high relief.

Mineral Association

It is a characteristic mineral of alkaline rocks where it occurs with nepheline, sphene etc. It is also present in peridotite with olivine and orthopyroxene.

Barite

Nature of the Phase

Essentially a pure phase in nature though extensive solid solution with celestite (SrSO$_4$) and anglesite (PbSO$_4$) has been synthesized as well as reported from natural samples. BaSO$_4$.

Crystallography

Orthorhombic, 2/m2/m2/m, Pnma.

$Z = 4$
$á = 1.636 - 1.637$
$â = 1.637 - 1.639$
$ã = 1.646 - 1.649$

Substitution of Sr decreases indices but that of Pb increases these values.

Plane Polarized Light

Form and Relief: Aggregate of small platy crystals, concentric cavity-filling etc. Relief strong.

Color and Pleochroism: Colorless and non-pleochroic.

Cleavage: Three sets of cleavages: {001} and {210} are perfect, {010} is good.

Under Crossed Polars

Anisotropism: Anisotropic.

Birefringence: Low, best seen on {010} form.

ä: 0.010 – 0.013.

Polarization Colors: First order yellow.

Extinction: Parallel to pinacoidal cleavage but symmetrical to {210} prismatic cleavage.

Orientation: á = z, â = y, ã = x (Fig. 16.3).

Indicatrix: Biaxial positive. Optic figure easily obtained on {100} form. Even basal section yields a good obtuse biaxial figure. The isogyres are reasonably broad with white background.

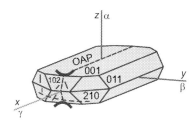

Fig. 16.3 Optic orientation of barite.

Non-silicates

$2V_a$: $36° - 40°$.

OAP: {010}.

Dispersion: Weak, $v > r$. Optic Axis dispersion.

Twinning: Not seen.

Zoning: Not seen.

Alteration

May alter to carbonates.

Distinguishing Characters

Mode of occurrence (vein and cavity filling), parallel extinction and cleavages. Gypsum has inclined extinction and negative relief.

Mineral Association

Gypsum, anhydrite and calcite are common associates.

Gypsum

Nature of the Phase

A pure phase. $CaSO_4.2H_2O$.

Crystallography

Monoclinic, 2/m, A2/a.

$$Z = 2$$
$$á = 1.519 - 1.521$$
$$â = 1.522 - 1.526$$
$$ã = 1.529 - 1.531$$

Plane Polarized Light

Form and Relief: Gypsum occurs in varied forms; it occurs as euhedral prismatic grains or plates, as an aggregate of fibres, or splinters. It also occurs as massive or earthy (Fig. 16.4).

Color and Pleochroism: Colorless and non-pleochroic.

Cleavage: Three sets of very good to perfect cleavages: {010}, {100}, {$\bar{1}$11}. Prismatic cleavage {$\bar{1}$11} shows up as two sets because of the presence of minor plane along {010} with angles of intersection as 42° and 138°.

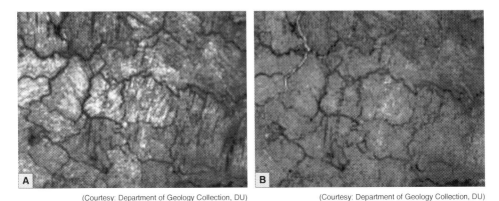

Fig. 16.4 Photomicrographs of gypsum. **A.** Plane polarized light. **B.** Crossed polars.

Under Crossed Polars

Anisotropism: Anisotropic.

Birefringence: Weak, best seen on {010}.

ã: 0.010.

Polarization Colors: First order pale to light yellow.

Extinction: Depends upon which cleavage set is being considered; with {010} the extinction is parallel but is inclined to {100} by 38°.

Orientation: á ∧ x = –15°, â = y, ã ∧ z = –52° (Fig. 16.5).

Indicatrix: Biaxial positive. Interference figure best seen on {001} section; isogyres are broad and diffused with few or non-isochromes.

2V$_a$: 58°.

OAP: {010}.

Sign of Elongation: Since both ã and á make high angle to the length of {010} face elongation can be positive or negative, but {100} cleavage traces on {010} are length fast.

Dispersion: Strong bisectrix as well as 2V dispersion. $r > v$.

Twinning: Very common on {100}.

Zoning: Not seen.

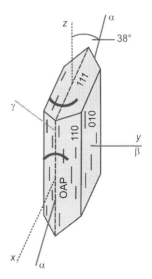

Fig. 16.5 Optic orientation of gypsum.

Non-silicates 347

Alteration

Not common; at times it alters to a mixture of carbonates and cryptocrystalline silica.

Distinguishing Characters

Three sets of cleavages, low negative relief, and low birefringence.

Mineral Association

Gypsum is a sedimentary mineral precipitated in a progressively drying basin. Hence, carbonates, sulfates and halides are common associates.

Anhydrite

Nature of the Phase

A pure phase. $CaSO_4$.

Crystallography

Orthorhombic, 2/m2/m2/m, Amma.

$$Z = 4$$
$$á = 1.569 - 1.574$$
$$â = 1.574 - 1.579$$
$$ã = 1.609 - 1.618$$

Plane Polarized Light

Form and Relief: Aggregates of rounded grains. Radiating fibres also common. Relief poor.

Color and Pleochroism: Colorless and non-pleochroic.

Cleavage: Three sets of very good mutually perpendicular cleavages: {010}, {100}, {001}.

Under Crossed Polars

Anisotropism: Anisotropic.

Birefringence: Moderate to strong; best seen on {100} form.

ä: 0.040.

Polarization Colors: High second order to low third order colors.

Extinction: Parallel to all cleavage traces.

Orientation: á = y, â = x, ã = z (Fig. 16.6).

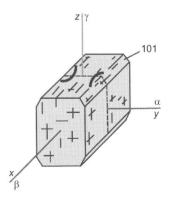

Indicatrix: Biaxial positive. The optic figure is best viewed normal to a basal section. Isogyres are sharp and thin with numerous isochromatic bands.

$2V_a$: 43°.

OAP: {100}.

Sign of Elongation: Normally it is not possible to predict which cleavage trace would be slow or fast direction because each of the three sets of cleavages contain two vibration directions and is normal to the third.

Fig. 16.6 Optic orientation of anhydrite.

Dispersion: Strong dispersion of the optic axis; v > r.

Twinning: Common. Twin law {011}; can be simple or polysynthetic.

Zoning: Not seen.

Alteration

Commonly alters to gypsum.

Distinguishing Characters

Three sets of cleavages and high birefringence.

Mineral Association

A sedimentary mineral. It precipitates in high salinity isolated shallow basin. Minerals characteristic of evaporite sequences like gypsum, barite, calcite etc. are common associate.

Calcite

Nature of the Phase

Nearly a pure phase. $CaCO_3$. At high temperature, it may incorporate small amount of Mg (most common), Fe, Mn, Zn, Sr and Ba.

Crystallography

Hexagonal, 2/m, R$\bar{3}$c.

$$Z_{hex} = 6 \qquad Z_{rh} = 2$$
$$ù = 1.658$$
$$å = 1.486$$

The above values are for pure calcite; when other cations substitute for Ca, these values will increase.

Plane Polarized Light

Form and Relief: In thin sections calcite appears as fine to coarse anhedral grains with a good relief that can be positive or negative depending upon the orientation of the grain with respect to the vibration direction of the polars. Note that ù is much higher than the refractive index of araldite, and å is much lower than that value. Because of such a contrast in the two refractive indices, calcite grain outline will show bold relief in one orientation of the grain; during rotation of the microscope stage the boundary would disappear owing to a particular direction of the crystal bearing the same index as the araldite, and then further rotation would bring a negative relief of the grain. It is known as the twinkling effect.

Color and Pleochroism: Colorless and non-pleochroic.

Cleavage: Perfect rhombohedral cleavage {101} (Fig. 16.7).

Under Crossed Polars

Anisotropism: Anisotropic but basal sections are isotropic.

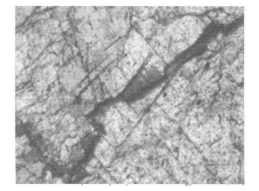

Fig. 16.7 Photomicrograph of calcite in plane polarized light.

Birefringence: Very high. It may even be apparent in basal sections because light rays become convergent to the optic axis. It is said that the extreme birefringence of calcite and other carbonates results from the coordination of CO_3^{2-} group which is planar equilateral triangle. The electric vector of light ray polarizes the three oxygens atoms facilitating the light to pass faster along the plane (the basal pinacoid) than normal to it, (*i.e.*, z-axis).

ä: 0.172.

Polarization Colors: Calcite produces fourth or higher order of interference colors. However, owing to strong convergence and abundant twin lamellae uniform grey and a shade of pink and green are visible.

Extinction: Since rhombohedral cleavages meet at a high angle the extinction angle may be symmetrical or inclined.

Indicatrix: Uniaxial negative. The interference figure shows sharp thin isogyres with numerous colored rings. In grains from deformed marbles, calcite may show anomalous biaxial behaviour with $2V_a$ up to 15°.

Sign of Elongation: Slow direction (å) is parallel to the shorter diagonal of the rhomb.

Dispersion: Very strong.

Twinning: Lamellar twinning on negative rhombohedron $\{01\bar{1}2\}$ invariably present (Figs. 16.8 and title page). Due to the inclined nature of the twin plane, overlap of optic orientation causes illumination of some degree in every position of the microscope rotation. An oblique plaid design of pale interference colors is a common result of twinning. These twins are glide twins resulting from application of stress to calcite grains, hence, are common in metamorphic rocks. Intersection of the cleavage rhomb with twin lamellae is along the long diagonal of the rhomb. Twinning on other laws also take place of which $\{0001\}$ is more common.

(Courtesy: Department of Geology Collection, DU)

Fig. 16.8 Photomicrograph of calcite in crossed polar.

Zoning: Not seen. However, trace element substitution may exhibit zoning in cathodoluminescence.

Alteration

Not seen.

Distinguishing Characters

It is easy to identify a hexagonal carbonate by virtue of its rhombohedral cleavage, birefringence and twinkling effect of its relief. Universal stage method can distinguish between different carbonate species, but, staining techniques and x-ray diffraction methods are easy and more definitive. Optically, a simple method to judge calcite from dolomite is to locate whether composition plane of the twin lamellae is parallel to short or long diagonal of the cleavage rhomb. In calcite it is parallel to long diagonal.

Non-silicates   **351**

Mineral Association

Calcite occurs in igneous, metamorphic, and sedimentary rocks, as well as in hydrothermally and metasomatically altered rocks. It occurs with quartz, nepheline, feldspars, several species of pyroxenes, amphiboles, sulfates, phosphates etc.

Dolomite

Nature of the Phase

Nearly a pure phase. (Ca, Mg)CO_3.

Crystallography

Hexagonal $\bar{3}$, R$\bar{3}$.

$$Z_{hex} = 3; Z_{hex} = 1$$
$$ù = 1.679$$
$$å = 1.510$$

Plane Polarized Light

Form and Relief: Commonly occurs as aggregates of small anhedral crystals. Relief variable; exhibits twinkling effect.

Color and Pleochroism: Colorless and non-pleochroic.

Cleavage: Typical rhombohedral cleavage.

Under Crossed Polars

Anisotropism: Anisotropic but basal sections are isotropic.

Birefringence: Very high with behaviour similar to that of calcite.

ä: 0.179 and increase if the substitution by Fe^{2+} becomes substantial to make the mineral ankerite.

Polarization Colors: High third order, but general appearance is similar to that of calcite.

Extinction: The angle between å and the {021} twin lamellae range between $20° - 40°$ provided the thin section is cut at a high angle to the twin lamellae.

Indicatrix: Uniaxial negative. Optic figure shows thin isogyres with numerous colored rings.

Sign of Elongation: Negative, though it is not easy to find a scalenohedron to decipher its length fast character.

Optical Mineralogy

352

Dispersion: Strong.

Twinning: Very common on $\{02\bar{2}1\}$, and on $\{11\bar{2}0\}$. When both twins are present the chances are that the grain under view is dolomite as opposed to calcite. However, in comparison to calcite, the twinning is not so ubiquitous.

Zoning: Not seen.

Alteration

Not seen.

Distinguishing Characters

Since the twin law is different from that of calcite, the composition plane in dolomite rhomb would parallel shorter diagonal. If the uncovered thin section is heated above 550°C for an hour or so then re-examine the dolomite will remain unaffected, but the carbonate was magnesite, it would dissociate into periclase+CO_2.

Mineral Association

Essentially a sedimentary mineral, it occurs with other carbonates like calcite, phosphates and sulfates.

Aragonite

Nature of the Phase

Nearly a pure phase. $CaCO_3$.

Crystallography

Orthorhombic, 2/m2/m2/m, Pmcn.

$$
\begin{aligned}
Z &= 4 \\
á &= 1.530 - 1.531 \\
â &= 1.680 - 1.681 \\
ã &= 1.685 - 1.686
\end{aligned}
$$

Plane Polarized Light

Form and Relief: Aragonite occurs in various forms, such as stalactitic, coral, but more commonly as small stout columnar aggregates. Relief is low negative to moderate positive with a small twinkling effect.

Color and Pleochroism: Colorless and non-pleochroic.

Cleavage: Faint {010}.

Under Crossed Polars

Anisotropism: Anisotropic.

Birefringence: Birefringence depends upon the section of the grain being viewed. Because of closeness in value between â and ã basal sections exhibit very low birefringence whereas the {100} pinacoidal section shows the strongest birefringence.

ã: 0.155.

Polarization Colors: Basal sections show low first order colors but vertical sections show third to fourth order colors.

Extinction: Parallel.

Orientation: á = z, â = x, ã = y (Fig. 16.9).

Indicatrix: Biaxial negative. Best seen on basal pinacoid numerous isochromes.

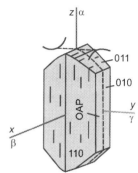

Fig. 16.9 Optic orientation of aragonite.

$2V_a$: 18°.

Sign of Elongation: Negative, cleavage traces are length fast.

Dispersion: Weak optic axis dispersion, v > r.

Twinning: Repeated twinning on {110} is very common. This causes its frequent habit of columnar grain with hexagonal forms; basal sections exhibit characteristic radial sectors.

Zoning: Not seen.

Alteration

Alters to calcite.

Distinguishing Characters

Higher relief than calcite, biaxial character and lack of rhombohedral cleavages are main distinguishing characters.

Mineral Association

Aragonite is restricted to low temperature Quaternary deposits associated with thermal springs etc. Also, reported from high pressure rocks like eclogites. Calcite, sulfates and other low temperature minerals are characteristically associated with aragonite.

Apatite

Nature of the Phase

Apatite is a group of several minerals all resembling closely in composition often forming solid solution amongst each other. The end-member formula of the most common member is $Ca_5(PO_4)_3(F, OH, Cl)$ and is referred to as the fluorapatite. Radio-elements are generally present in trace amount.

Crystallography

Hexagonal, $6/m$, $C6_3/m$.

$$Z = 2$$
$$ù = 1.603 - 1.667$$
$$å = 1.619 - 1.665$$

The above values show large range for each index because of large multisite variation elements in apatite formula. The correlation between refractive indices and chemical composition is not established.

Plane Polarized Light

Form and Relief: Rectangular prismatic and hexagonal basal sections with prominent relief (Fig. 16.10).

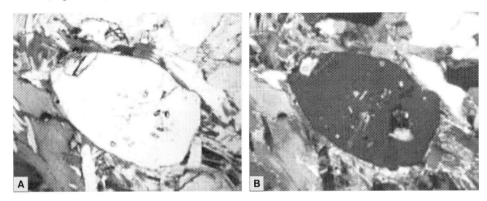

Fig. 16.10 Photomicrographs of apatite. **A.** Plane polarized light; **B.** Crossed polars.

Non-silicates   355

Color and Pleochroism: Essentially colorless and non-pleochroic; in rare case pale colored and mildly pleochroic.

Cleavage: Normally not seen.

Under Crossed Polars

Anisotropism: Anisotropic but basal sections are isotropic.

Birefringence: Low to moderate; best seen on prismatic sections.

ä: $0.001 - 0.007$. The birefringence also depends upon the composition.

Polarization Colors: Low first order grey to yellow.

Extinction: Parallel.

Indicatrix: Uniaxial negative; diffused isogyres on a first order grey field. The interference figure best seen on the basal pinacoid. There are examples of apatite showing biaxial figure.

Sign of Elongation: Negative (length fast).

Dispersion: Moderate.

Twinning: Not seen.

Zoning: Present though not so common.

Alteration

Not present.

Distinguishing Characters

High relief and low birefringence coupled with uniaxial characters.

Mineral Association

It is a common igneous mineral and an important accessory in most igneous rocks. It is also a common detrital mineral. Apatite is quite often present in carbonate rocks. Also, quite common in metamorphic rocks.

Monazite

Nature of the Phase

Monazite composition is highly variable as elemental proportions of rare earth elements may vary; in addition, a number of substitutions also takes place (Ce, La, Th)PO_4.

Crystallography

Monoclinic, 2/m, P2$_1$/n.

$$
\begin{aligned}
Z &= 4 \\
á &= 1.770 - 1.800 \\
â &= 1.777 - 1.801 \\
ã &= 1.828 - 1.851
\end{aligned}
$$

Plane Polarized Light

Form and Relief: Small euhedral crystals elongated along *b*-axis with very high relief (Fig. 16.11).

Color and Pleochroism: Grains exhibit grey to pale yellow color. Weakly pleochroic. â > á ≅ ã. á = light yellow, â ≅ ã = dark yellow.

Cleavage: Two sets of cleavages: {100} distinct, and {010} poor intersecting at 90°.

(Courtesy: N.C. Pant)

Fig. 16.11 Photomicrograph of a small inclusion of

Under Crossed Polars

Anisotropism: Anisotropic.

Birefringence: High; best seen on {100} sections.

ä: 0.045 – 0.075.

Polarization Colors: High third order, may even range up to lower fourth order (Fig. 16.12).

(Courtesy: N.C. Pant)

Fig. 16.12 Photomicrograph of an euhedral monazite in crossed polars.

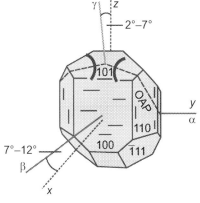

Fig. 16.13 Optic orientation of monazite.

Extinction: Inclined extinction with a small extinction angle; maximum value is obtained on {010} face up to 7° with {100} cleavage.

Orientation: á = y, â ∧ x = +12° to + 7°, ã ∧ z = –2° to –7° (Fig. 16.13).

Non-silicates

Indicatrix: Biaxial positive; optic figure on basal pinacoid is well-centred with sharp isogyres and numerous isochromatic bands.

$2V_a$: Small angle going up to 19°.

Sign of Elongation: Grains elongated along y-direction are length fast, although monazite is positive.

OAP: Parallel to {100}.

Dispersion: Weak, $v > r$.

Twinning: Simple twins common on {101}; rare lamellar on {001}.

Zoning: Not seen.

Alteration

Not common.

Distinguishing Characters

High positive relief, high birefringence, and weak pleochroism are characteristic properties. Zircon similar though uniaxial.

Mineral Association

It is an important accessory mineral in silica rich and intermediate rocks. Quartz and feldspar are common associates. It also occurs in metamorphic rocks especially in metapelites.

Xenotime

Nature of the Phase

Essentially a pure phase. YPO_4. However, yttrium is easily substituted by several elements.

Crystallography

Tetragonal.

Z	
ù =	1.719 – 1.724
å =	1.816 – 1.827

358   Optical Mineralogy

Plane Polarized Light

Form and Relief: Most common habit is tiny crystals resembling those of zircon. Very prominent relief.

Color and Pleochroism: Colorless to pale yellow. Pleochroic, ù = pale yellow, å = yellow.

Cleavage: Good prismatic cleavage {110} intersecting at 90°.

Under Crossed Polars

Anisotropism: Anisotropic. Basal sections are isotropic.

Birefringence: Very high. Best seen in prismatic sections.

ä: $0.095 - 0.107$.

Polarization Colors: Very high order of grey interference colors; even basal sections may show polarization colors.

Extinction: Parallel in prismatic section; symmetrical to {110} cleavage in oblique sections.

Indicatrix: Uniaxial positive. Basal section reveals a centred figure with sharp isogyres and numerous isochromatic rings.

Sign of Elongation: Elongation is positive (length slow).

Zoning: May be present.

Alteration

Not present.

Distinguishing Characters

It can be easily confused with zircon but has lower relief and higher birefringence.

Mineral Association

Like zircon, it has wide occurrence as an accessory mineral in granites and syenites, schists and gneisses.

Fluorite

Nature of the Phase

Essentially a pure phase. CaF_2.

Non-silicates

Crystallography

Isometric, 4/mm, Fm3m.

$$Z = 4$$
$$n = 1.433 - 1.435$$

Plane Polarized Light

Form and Relief: Aggregates of medium to coarse-grained subhedral to euhedral grains with high negative relief.

Color and Pleochroism: Commonly colorless and non-pleochroic.

Cleavage: Perfect octahedral cleavage.

Under Crossed Polars

Anisotropism: Isotropic.

Twinning: Very commonly twinned on {111} but twinning is not obvious under microscope due to its isotropic nature.

Dispersion: Weak.

Zoning: Not seen.

Alteration

Easily altered to chalcedony, carbonates and Fe/Mn oxides.

Distinguishing Characters

Strong negative relief and octahedral cleavage along with isotropic character allow its easy identification.

Mineral Association

Fluorite represents a phase of hydrothermal alteration, hence is restricted to veins, fracture and cavity filling. Common associated minerals are calcite, quartz, tourmaline, topaz etc.

Halite

Nature of the Phase

Essentially a pure phase. NaCl.

Crystallography

Isometric, 4/mm, Fm3m.

$$Z = 4$$
$$n = 1.544$$

Plane Polarized Light

Form and Relief: Aggregates of rounded grains with very low relief.

Color and Pleochroism: Colorless and non-pleochroic.

Cleavage: Perfect cubic cleavage.

Under Crossed Polars

Anisotropism: Anisotropic.

Twinning: Not seen.

Zoning: Not seen.

Alteration

Highly solubles, hence a cast cavity may be left in its place.

Distinguishing Characters

Perfect cubic cleavage and low relief along with isotropic character.

Mineral Association

Common evaporite mineral, halite occurs with other evaporite minerals such as gypsum, anhydrite, sylvite etc.

■■■

Appendix 1

Relief	Birefringence						
	Low		Medium		High		
Low	Stilbite Heulandite Thomsonite G ypsum	(010), P	Cordierite	(010), D			
	Chlorite	(001), P					
Low to moderate					Prehnite	(001), D	
					Talc Muscovite Lepidolite	(001), P	
Moderate to high	Topaz Clinozoisite	(001), P	Prehnite	(001), D	Phlogopite Biotite		
	Chloritoid		Silimanite	(100), P	Stilpnomelane Epidote Piedmontite	(001), P	
	Zoisite	(010),					
	Staurolite	P			Monazite	Parting (001)	
	Corundum	(0001)					

© The Author(s) 2023
P. K. Verma, *Optical Mineralogy*,
https://doi.org/10.1007/978-3-031-40765-9

Table 1

Relief	Birefringence					
	Low		**Medium**		**High**	
Low	Orthoclase Sanidine Anorthoclase Microcline Albite Plagioclase	(001), P (010), D	Natrolite Scapolite	(110), D		
Moderate to high	Enstatite	(210), P	Tremolite actinolite	(110), P	Diopside Grunerite	(110), P
	Andalusite	(110), D	Glaucophane Hornblende		Pigeonite Aegirine	
	Riebeckite	(110), P	Cummingtonite Augite Aegirine-augite Spodumene Anthophyllite Enstatite	(210), P	Rutile	(110), D

Table 2

Relief	Birefringence					
Low	Halite	(100), P	Scapolite	(100), P; (110), D	Anhydrite	(001); (010); (001) P
			Celestite	(001), P (110), D		
Moderate to high relief	Corundum	(1011), partings	Barite	(001); (110), P		
			Lawsonite	(010), P; (110), D		
			Kyanite	(110), (010), P; (001) parting		
Extreme Relief	Periclase Perovskite	(100), P			Rutile	(100); (110), D

Appendix 2

Isotropic Minerals		
S.No.	R. I.	Mineral Name
1.	1.434	Fluorite
2.	1.483 – 1.487	Sodalite
3.	1.544	Halite
4.	1.72 – 1.78	Spinel
5.	1.736 – 1.763	Grossular
6.	1.738 – 1.760	Periclase
7.	1.741 – 1.760	Pyrope
8.	1.778 – 1.815	Almendine
9.	1.792 – 1.820	Spessartine
10.	1.838 – 1.870	Uvarovite
11.	1.857 – 1.887	Andradite
12.	2.34 – 2.38	Perovskite

Uniaxial Minerals		
S.No.	ω	ε	Name
1.	1.658	1.486	Calcite
2.	1.680 – 1.716	1.500 – 1.526	Dolomite
3.	1.700 – 1.726	1.509 – 1.527	Magnesite
4.	1530 – 1.547	1.527 – 1.543	Nephaline
5.	1.550 – 1.607	1.540 – 1.571	Scapolite
6.	1.5442	1.5533	Quartz
7.	1.568 – 1.598	1.564 – 1.590	Beryl
8.		1.628 – 1.658	Tourmaline
9.	1.633 – 1.655		Apatite
		1.613 – 1.628	Dravite (tourmaline)
10.	1.635 – 1.655		Elbaite (tourmaline)
11.	1.652 – 1.698		Schorlite (tourmaline)
12.	1.705 – 1.732		Vesuvianite
13.	1.767 – 1.772	1.759 – 1.763	Corundum
14.	1.830 – 1.875		Siderite
15.	1.985 – 1.931	1.985 – 1.993	Zircon
16.	2.603 – 2.616	2.889 – 2.903	Rutile

© The Author(s) 2023
P. K. Verma, *Optical Mineralogy*,
https://doi.org/10.1007/978-3-031-40765-9

Biaxial Minerals

α	β	γ	Name
1.469		1.473	Tridymite
1.473 – 1.480	1.485 – 1.493		Natrolite
1.478 – 1.485	1.480 – 1.490		Chabazite
1.484		1.487	Cristobalite
1.493 – 1.546	1.517 – 1.557		Chrysotile
1.494 – 1.500			Stilbite
1.496 – 1.499			Neulandite
1.508		1.509	Leucite
1.517 – 1.520	1.524 – 1.526		Sanidine
1.518		1.526	Orthoclase
1.518 – 1.522	1.525 – 1.530		Microcline
1.52		1.529	Gypsum
1.522 – 1.536	1.527 – 1.541		Anorthoclase
1.525 – 1.532	1.536 – 1.541		Albite
1.53			Aragonite
1.532 – 1.545	1.541 – 1.552		Oligoclase
1.532 – 1.552	1.539 – 1.570		Cordierite
1.538 – 1.545	1.575 – 1.590		Talc
1.541 – 1.579	1.574 – 1.5638		Biotite
1.552 – 1.562			Andesine
1.551 – 1.562	1.598 – 1.606		Phlogopite
1.555 – 1.563	1.562 – 1.571		Labradorite
1.555 – 1.564	1.562 – 1.573		Antigorite
1.556 – 1.570	1.593 – 1.611		Muscovite
1.56		1.605	Lepidolite
1.563 – 1.571	1.571 – 1.582		Bytownite
1.57	1.614		Anhydrite
1.571 – 1.575	1.582 – 1.588		Anorthorite
1.571 – 1.588	1.576 – 1.597		Clinochlore
1.575 – 1.582			Pennine
1.598 – 1.652	1.623 – 1.676		Anthophyllite
1.600 – 1.628	1.625 – 1.655		Tremolite-Actinolite
1.607 – 1.629	1.617 – 1.638		Topaz
1.612 – 634			Stilpnomelane
1.614 – 1.675	1.633 – 1.701		Hornblende
1.615 – 1.635	1.645 – 1.665		Prehnite
1.62		1.634	Wollastonite
1.621 – 1.655	1.639 – 1.668		Glaucophane
1.629 – 1.640	1.639 – 1.647		Andalusite
1.635 – 1.640	1.670 – 1.680		Forsterite
1.648		1.648	Barite
1.639 – 1.657	1.664 – 1.686		Cummingtonite
1.64 – 1.77	1.66 – 1.80		Allanite

α	β	γ	Name
1.641 – 1.651	1.655 – 1.669		Monticellite
1.642		1.654	Mullite
1.650 – 1.665	1.658 – 1.674		Enstatite
1.650 – 1.698			Diopside
1.651 – 1.668	1.677 – 1.681		Spodumine
1.651 – 1.681			Olivine
1.655 – 1.666	1.667 – 1.688		Jadeite
1.657 – 1.661	1.677 – 1.684		Sillimanite
1.657 – 1.663			Grunerite
1.665			Lawsonite
1.673 – 1.715			Ferrienstatite
1.680 – 1.718			Pigeonite

Appendix 3

Mineral	Optic Character	Optic Sign	Birefringence	2V for Biaxial Mineral
Perovskite			0.00 – 0.002	
Antigorite	biaxial	negative	0.00 – 0.003	
Leucite			0.001	
Penine	biaxial	positive	0.001 – 0.004	
Chabazite	biaxial	positive	0.002 – 0.010	
Cristobalite			0.003	
Apatite	uniaxial	negative	0.003 – 0.004	
Nepheline	uniaxial	negative	0.003 – 0.004	
Tridymite	biaxial	positive	0.004	
Riebeckite	biaxial		0.004	
Vesuvianite	uniaxial	negative	0.004 – 0.006	
Beryl	uniaxial	negative	0.004 – 0.006	
Clinochlore	biaxial	positive	0.004 – 0.011	
anorthoclase	biaxial	negative	0.005 – 0.007	
Clinozoisite	biaxial	positive	0.005 – 0.011	
Stilbite	biaxial	negative	0.006 – 0.008	
Zoisite	biaxial		0.006 – 0.018	
Sanindine	biaxial		0.007	
Heulandite			0.007	
Andesine	biaxial		0.007	
Microcline	biaxial		0.007	
Labradorite	biaxial		0.007 – 0.008	
Antigorite	biaxial		0.007 – 0.009	
Oligoclase	biaxial		0.007 – 0.011	
Andalusite	biaxial		0.007 – 0.011	
Cordierite	biaxial	positive	0.007 – 0.011	
Orthoclase	biaxial		0.008	
Corundum	uniaxial	negative	0.008 – 0.009	
Enstatite	biaxial	positive	0.008 – 0.009	
Bytownite	biaxial		0.008 – 0.011	
Celestite	uniaxial	positive	0.009	
Gypsum	biaxial		0.009	
Quartz	uniaxial	positive	0.009	
Topaz	biaxial	positive	0.009 – 0.010	
Albite	biaxial		0.009 – 0.011	
Staurolite	biaxial		0.010 – 0.015	
Ferienstatite	biaxial	negative	0.010 – 0.016	

© The Author(s) 2023
P. K. Verma, *Optical Mineralogy*,
https://doi.org/10.1007/978-3-031-40765-9

Mineral	Optic Character	Optic Sign	Birefringence	2V for Biaxial Mineral
Allanite	biaxial		0.01 – 0.03	
Scapolite	uniaxial		0.010 – 0.036	
Anorthite	biaxial		0.011 – 0.013	
Chrysotile	biaxial	negative	0.011 – 0.014	
Barite	biaxial		0.012	
Mullite			0.012	
Natrollite			0.012 – 0.013	
Jadeite	biaxial		0.012 – 0.023	
Chloritoid	biaxial		0.013 – 0.016	
Glaucophane	biaxial		0.013 – 0.016	
Wollastonite	biaxial		0.014	
Monticellite	biaxial		0.014 – 0.018	
Epidote	biaxial	negative	0.014 – 0.045	
Elbaite(tourmaline)	uniaxial	negative	0.015 – 0.023	
Kyanite	biaxial	negative	0.016	
Anthophyllite	biaxial		0.016 – 0.025	
Hedenbergite	biaxial	negative	0.018 – 0.0019	
Lawsonite			0.019	
Dravite	uniaxial	negative	0.019 – 0.025	
Hornblende	biaxial		0.019 – 0.026	
Sillimanite	biaxial	positive	0.020 – 0.023	
Prehnite			0.020 – 0.033	
Augite	biaxial		0.021 – 0.025	
Pigeonite	biaxial		0.021 – 0.033	
Tremolite	biaxial		0.022 – 0.0027	
Schorlite	uniaxial	negative	0.022 – 0.040	
Cummingtonite	biaxial		0.025 – 0.029	
Diopside	biaxial	positive	0.029 – 0.031	
Aegirine-augite	biaxial		0.029 – 0.037	
Talc	biaxial	negative	0.03 – 0.035	
Stilpnomelane			0.030 – 0.119	
Biotite	biaxial	negative	0.033 – 0.059	
Forsterite	biaxial	positive	0.035 – 0.040	
Olivine	biaxial	negative	0.037 – 0.041	
Muscovite	biaxial	negative	0.037 – 0.041	
Aegirine-augite	biaxial		0.037 – 0.059	
Fayalite	biaxial	negative	0.042 – 0.051	
Grunerite	biaxial		0.042 – 0.054	
Anhydrite	biaxial	positive	0.044	
Phlogopite			0.044 – 0.047	
Monazite			0.049 – 0.051	
Zircon	uniaxial	positive	0.060 – 0.062	
Piedmontite	biaxial		0.061 – 0.082	
Sphene			0.092 – 0.141	
Aragonite	biaxial		0.156	
Calcite	uniaxial	negative	0.172	
Dolomite	uniaxial		0.180 – 0.190	
Magnesite	uniaxial		0.191 – 0.199	
Rutile	uniaxial		0.286 – 0.287	

Appendix 4

Common liquids for refractive index determination

Liquid	n	°C	Dn/dt	Dispersion
Acetone	1.359	19.4	–	Slight
Ethyl butyrate	1.362	20	–	Slight
Ethyl bromide	1.424	20	–	–
Methyl thiocyanate	1.446	25	0.00054	–
Benzene	1.501	20	–	–
Monochlorobenzene	1.525	20	–	–
Dimethylaniline	1.559	20	–	–
Bromoform	1.598	19.0	−0.0006	–
Monoiodobenzene	1.621	18.5	–	–
Potassium mercuric iodide solution	< 1.71	–	–	–
Methylene iodide	1.74	–	−0.0007	strong

© The Author(s) 2023
P. K. Verma, *Optical Mineralogy*,
https://doi.org/10.1007/978-3-031-40765-9

Printed in the United States
by Baker & Taylor Publisher Services